Pitman Research Notes in Mathematics Series

Submission of proposals for consideration

Suggestions for publication, in the form of outlines and representative samples, are invited by the Editorial Board for assessment. Intending authors should approach one of the main editors or another member of the Editorial Board, citing the relevant AMS subject classifications. Alternatively, outlines may be sent directly to the publisher's offices. Refereeing is by members of the board and other mathematical authorities in the topic concerned, throughout the world.

Preparation of accepted manuscripts

On acceptance of a proposal, the publisher will supply full instructions for the preparation of manuscripts in a form suitable for direct photo-lithographic reproduction. Specially printed grid sheets can be provided and a contribution is offered by the publisher towards the cost of typing. Word processor output, subject to the publisher's approval, is also acceptable.

Illustrations should be prepared by the authors, ready for direct reproduction without further improvement. The use of hand-drawn symbols should be avoided wherever possible, in order to maintain maximum clarity of the text.

The publisher will be pleased to give any guidance necessary during the preparation of a typescript, and will be happy to answer any queries.

Important note

In order to avoid later retyping, intending authors are strongly urged not to begin final preparation of a typescript before receiving the publisher's guidelines. In this way it is hoped to preserve the uniform appearance of the series.

Longman Scientific & Technical
Longman House
Burnt Mill
Harlow, Essex, CM20 2JE
UK
(Telephone (0279) 426721)

Martin Hanke

Universität Karlsruhe, Germany

Conjugate gradient type methods for ill-posed problems

CRC Press

Taylor & Francis Group

Boca Raton London New York

CRC Press is an imprint of the
Taylor & Francis Group, an **informa** business

First published 1995 by Longman Group Limited

Published 2019 by Chapman & Hall\CRC
Taylor & Francis Group
6000 Broken Sound Parkway NW, Suite 300
Boca Raton, FL 33487-2742

© 1995 by Taylor & Francis Group, LLC
CRC Press is an imprint of Taylor & Francis Group, an Informa business

First issued in paperback 2019

No claim to original U.S. Government works

ISBN 13: 978-0-367-44911-7 (pbk)
ISBN 13: 978-0-582-27370-2 (hbk)
ISSN 0269-3664

Visit the Taylor & Francis Web site at
http://www.taylorandfrancis.com

and the CRC Press Web site at
http://www.crcpress.com

Copublished in the United States with
John Wiley & Sons Inc.

AMS Subject Classifications: (Main) 65-02, 65J20, 65J10
(Subsidiary) 42C05, 65F10

British Library Cataloguing in Publication Data

A catalogue record for this book is
available from the British Library

Library of Congress Cataloging-in-Publication Data

A catalog record for this book is available

Table of Contents

1. Preface

A key issue in many scientific problems is the solution of linear equations

$$Tx = y. \tag{1.1}$$

The unknown x in (1.1) may either be the solution of the entire problem or some intermediate quantity, as is the case, for example, when solving a nonlinear problem by Newton-type methods.

While T is typically a bounded operator between Hilbert spaces $\mathcal{X}$ and $\mathcal{Y}$, the dependency $y \mapsto x$ is not always continuous, in which case (1.1) is called *ill-posed*. This notion, which goes back to lectures by HADAMARD [33] from 1923, originated from the philosophy that a solution of such a problem cannot have a well defined physical sense. In the meantime, however, several important applications from natural sciences - so-called *inverse problems* - have made the study of ill-posed problems an essential task of applied mathematics. This is illustrated, for example, by the rapidly growing number of publications in the field, cf., e.g., the conference proceedings [72, 83, 79, 2, 66, 17], to name just a few. In most of these applications the existence of a solution follows from physical considerations; however, x does not need to be uniquely defined by (1.1), in case T has a nontrivial nullspace $\mathcal{N}(T)$. Therefore one typically searches the unique solution $x = T^\dagger y$ in $\mathcal{X}$ of minimal norm; the linear map $T^\dagger$ is the generalized inverse of T. Notice that ill-posedness is equivalent to $T^\dagger$ being unbounded.

The most often cited class of examples for ill-posed problems constitutes of Fredholm integral equations of the first kind with non-degenerate kernel function. The corresponding integral operator T is compact, and its singular values cluster at the origin. From the minmax principle for the singular values one easily concludes that $T^\dagger$ is unbounded.

In practice the data y are rarely given exactly; instead, due to measurement errors, modelling errors and such, there is some inherent noise in the data. In view of the discontinuity of $T^\dagger$, straightforward inversion of (1.1) is therefore not recommended. More sophisticated algorithms for solving ill-posed problems go back to the early sixties, when TIKHONOV [77] derived the concept of *regularization*. Roughly speaking, the idea behind regularization is to balance approximation error and propagated data error; in other words, in order to prevent unbounded magnification of the data error, one has to put up with a limited degree of approximation. This degree of approximation - let it be measured by some number $\varepsilon > 0$ - is a free parameter, the regularization parameter. If $\{y^\delta\}_{\delta>0}$ denotes a sequence of perturbed data with $\|y - y^\delta\| \leq \delta$, then one would like to choose ε in such a way that

1

$$x_\epsilon = x_\epsilon(T, y^\delta, \delta) \longrightarrow x = T^\dagger y \qquad \text{as } \delta \to 0.$$

In principle, every scheme of numerical approximation may serve as underlying rule for constructing x_ϵ: typically, $x_\epsilon = R_\epsilon y^\delta$, where $\{R_\epsilon\}_{\epsilon > 0}$ are (possibly nonlinear) operators converging pointwise to $T^\dagger$ on the range $\mathcal{R}(T)$ as $\epsilon \to 0$. For example, in Tikhonov regularization,

$$R_\epsilon = (T^*T + \epsilon I)^{-1} T^*, \qquad \epsilon > 0, \tag{1.2}$$

where $T^* : \mathcal{Y} \to \mathcal{X}$ is the adjoint operator of T. In other words, x_ϵ is the solution of the damped normal equation

$$(T^*T + \epsilon I)x_\epsilon = T^*y^\delta.$$

The crux of the matter is the practical choice of the regularization parameter for a fixed data set y^δ. The optimal parameter, i.e., the one which minimizes $\|R_\epsilon y^\delta - T^\dagger y\|$, is impossible to determine since the exact solution is not known; instead one may ask for asymptotically optimal rate of convergence of $x_\epsilon \to T^\dagger y$ as $\delta \to 0$. Although this rate of convergence is arbitrarily slow in general, it can be estimated given certain a priori information about the exact solution. A typical assumption used in the mathematical literature is stated in Assumption 3.6 below. Depending on a parameter $\mu > 0$, this assumption enables convergence rates

$$\|x_\epsilon(T, y^\delta, \delta) - T^\dagger y\| = O(\delta^{\mu/\mu+1}), \qquad \delta \to 0,$$

which are best possible in some uniform sense. A regularization method which realizes the above convergence rate is called *order-optimal* (for μ). The interested reader will find more details on the basic philosophies and theoretical grounds of regularization methods in [30, 82, 54, 16].

For a long time, Tikhonov regularization has been the one and only alternative for solving ill-posed problems. However, its numerical implementation can be rather expensive, in particular when it becomes necessary to compute approximations for several different regularization parameters. Note that a linear system with a selfadjoint, positive definite matrix has to be inverted for each choice of ϵ. For this reason direct approaches for solving the discrete problem usually fail for higher dimensional applications from natural sciences for the lack of computer time and storage restrictions. Instead, the damped normal equation is often solved by some iterative technique, e.g., by the method of conjugate gradients.

On the other hand, iterative methods have an inherent regularizing property when applied straight to problem (1.1). Although the iterates $x_k = x_k(T, y^\delta, \delta)$ diverge as $k \to \infty$ when the perturbed data y^δ do not belong to the domain of $T^\dagger$, the data error propagation remains limited in the beginning of the iteration. The quality of the optimal approximation therefore depends on how many iterative steps can be

2

performed until the iterates turn to diverge. This phenomenon - convergence in the beginning of the iteration, divergence eventually - has been called *semiconvergence* in [57]. The idea is now to stop the iteration at about the point where divergence sets in. In other words, the iteration count is the regularization parameter which remains to be controlled by an appropriate stopping rule. Notice how approximation error and propagated data error are balanced in this way.

NEMIROVSKII [58] was the first to come up with a rigorous treatment of the regularizing properties of the conjugate gradient method applied to $T^*Tx = T^*y$; he suggested the discrepancy principle (cf. Stopping Rule 3.10) for determining an appropriate regularization parameter k. Chapter 3 is essentially following his analysis to prove order-optimality of this stopping rule. When T itself is selfadjoint and semidefinite one might prefer to avoid the detour via the normal equation, and apply the conjugate gradient method straight to the original equation (1.1). NEMIROVSKII's technique can not be used in this case, nor does it apply to the minimal error method proposed by KING [47] (a variant of conjugate gradients for the normal equation). Chapter 4 presents an adequate stopping rule (different from the discrepancy principle) for these two situations, and analyzes its convergence properties.

The basic tools to be used throughout this book are the same as in [58], namely elementary properties of orthogonal polynomials over $\mathbb{R}$; the corresponding orthogonality measures are induced by the spectral decomposition of the underlying selfadjoint operator T or TT^*, respectively. The major emphasis is on the case when T is selfadjoint and semidefinite; however, the analysis is not restricted to this special case: all results have their natural analog for the alternative algorithms based on the normal equation as will be outlined in Section 2.3 and clarified in several remarks scattered throughout the entire text.

The brief outline of the following chapters is as follows. Basic facts about conjugate gradient type methods including the most important algorithms are presented in Chapter 2. Chapters 3 and 4 derive the regularizing properties of a family of conjugate gradient type methods for semidefinite problems: those which can be treated by NEMIROVSKII's techniques are considered in Chapter 3; the conjugate gradient method applied to (1.1) and the minimal error method for the normal equation (which cannot be analyzed in this way) are studied in Chapter 4. Each of these chapters deals with the following three major questions:

- under what conditions converges/diverges the iteration as the iteration index goes to infinity ?

- given the noise level δ in the data, how can the stopping index be chosen such that the approximations of $T^\dagger y$ are order-optimal ?

- are there heuristic stopping rules for the case that no information about the noise level δ is known ?

Estimates for the number of iterations until the stopping criterium is met are given in Chapter 5. Section 5.3 contains numerical experiments with an ill-posed deconvolution problem arising in image reconstruction. Finally, Chapter 6 deals with an extension of the former results to selfadjoint indefinite problems. A number of modifications are necessary in this context, but the major conclusions remain valid. Nevertheless, the theory for indefinite problems is not as complete as for the semidefinite case. It should be stressed that the extension to indefinite problems is not for purely academic reasons; two applications where the indefinite algorithms perform significantly better are presented in Section 6.7.

Bibliographical notes concerning the presented results and references to related works are summarized at the end of each chapter. At this point it should be emphasized that much of the work on conjugate gradient type methods for ill-posed problems originated from the Russian literature; although I have tried to gather as many Russian papers and monographs as possible, the list of references may not include all important contributions. I would like to thank Dr. Robert Plato who provided me with a number of original sources.

This book is an outgrowth of my habilitation thesis "Regularization of ill-posed problems by conjugate gradient type methods", but it contains a significant portion of additional material including, in particular, the whole chapter on indefinite problems. I want to use the occasion to acknowledge the aids I received from Professor Dr. W. Niethammer in particular, but also from my colleagues during the years of my habilitation at the University of Karlsruhe. Furthermore, I am grateful to Professor Dr. H. W. Engl and Professor C. W. Groetsch for their encouragement and support to publish this book.

Notation

Some notation has already been introduced: $\| \cdot \|$ and $\langle \cdot, \cdot \rangle$ denote norm and inner product in $\mathcal{X}$ and $\mathcal{Y}$, respectively; which of the two spaces is concerned is always clear from the context. Besides, the maximum norm over the interval $[a, b]$ is denoted by $\| \cdot \|_{[a,b]}$. The operator T has null space $\mathcal{N}(T)$ and range $\mathcal{R}(T)$, P denotes the orthoprojector onto the closure of $\mathcal{R}(T)$. $T^\dagger$ is the (Moore-Penrose) generalized inverse of T; for a formal definition of $T^\dagger$ and a discussion of its continuity, cf. [29]. The domain of $T^\dagger$ is denoted by $\mathcal{D}(T^\dagger)$. If not said otherwise, T is assumed to be selfadjoint, positive semidefinite with its spectrum contained in $[0, 1]$; obviously, (1.1) can always be rescaled to guarantee $\|T\| \leq 1$. The spectral decomposition of T (cf. [71]) defines a spectral family of orthogonal projectors $\{E_\lambda\}$; as usual, E_λ is defined to be continuous from the right. If $\alpha(\lambda)$ is a nondecreasing distribution function with a jump at $\lambda = 0$ then the integrals $\int_0 \ldots d\alpha(\lambda)$ and $\int^0 \ldots d\alpha(\lambda)$ include the contribution coming from this jump at the origin; otherwise, the notations $\int_{0+}$ and $\int^{0-}$ will be used.

The following list refers to the definition of other symbols which have a fixed meaning throughout the entire text:

$\mathcal{K}_k$	Section 2.1	$\mathcal{E}_k$	Section 6.2
Π_k, Π_k^0	Section 2.1	$\sum'$	(6.14)
Π_k^{00}	(6.4)	κ	(2.12)
p_k, q_{k-1}	(2.2)	$\pi_{k,n}$	Proposition 2.5
$[\cdot, \cdot]_n$	(2.5)	$\theta_{k,n}$	Proposition 2.8
$p_k^{[n]}$	(2.6)	w, ω	Assumption 3.6
$u_k^{[n]}$	(6.8)	μ	Assumption 3.6
$\lambda_{j,k}^{[n]}$	(2.19)	τ	Stopping Rules 3.10, 4.7
$\lambda_{+,k}, \lambda_{-,k}$	Section 6.2	ρ_k	(3.12), (6.37)
$\lambda_{0,k}$	(6.24)	ρ_*, δ_*	Theorems 3.14, 4.11
$E_{\mathcal{I}}$	(6.26)	ϱ_k	Corollary 6.2
$T_\varepsilon^\dagger$	(2.15)	η_k	(3.25), (4.26), (6.55)
R_k	Proposition 2.10	γ	(3.29)
x_k^δ	Section 3.3	ϑ_k	(4.11)

Many results in this work have an asymptotic nature. Besides the usual Landau symbols $o(\cdot)$ and $O(\cdot)$, two notations will be used. Let $\{a_k\}$ and $\{b_k\}$ be nonnegative sequences: then

$$a_k \sim b_k, \qquad k \to \infty,$$

if and only if there are positive constants c_1 and c_2 with $c_1 b_k \leq a_k \leq c_2 b_k$ for all k sufficiently large; on the other hand,

5

$$a_k \doteq b_k, \qquad k \to \infty,$$

if and only if $b_k \neq 0$ for k sufficiently large, and $a_k/b_k \to 1$ as $k \to \infty$. Further on, the letter c always denotes a generic positive constant, i.e., c may be attached to different constants at different places, but always independent of the variables in question.

A common feature in the theory of ill-posed problems is the occurrence of fractional powers as, e.g., in $\delta^{1/2\mu+2}$. For notational convenience (and for the ease of the reader) paranthesis around the denominator are omitted throughout the entire text: everything following "/" belongs to the denominator. In other words, the correct notation for the above formula would have been $\delta^{1/(2\mu+2)}$.

2. Conjugate Gradient Type Methods

This preliminary chapter starts with a general description of Krylov subspace methods (also called polynomial iteration methods) and introduces the particular definition of what will further on be called conjugate gradient type methods; the most important algorithms are presented in slightly more detail. Conjugate gradient type methods may be viewed as optimization techniques or as projection methods. Alternatively, they can be studied from an orthogonal polynomial point of view, and this is the framework which is chosen in the following chapters. As can be seen in Section 2.4, many elementary properties of real orthogonal polynomials have interesting implications on conjugate gradient type methods. One consequence, for example, is a way of implementing two "adjacent" conjugate gradient type methods with essentially the same costs as implementing only one scheme; this is shown in Section 2.5. The final section of this chapter deals with the stability of the numerical algorithms. One aspect is the loss of orthogonality due to finite precision arithmetic, and its impact on the performance of the methods; another point is the sensitivity with respect to data perturbations after a fixed number of steps. The operator which maps the right-hand side onto the corresponding iterate is nonlinear and may be discontinuous, but discontinuity is restricted to a small set, i.e., a set of first category. Except for Section 2.3, T is always assumed to be selfadjoint, semidefinite, with its spectrum contained in $[0, 1]$.

2.1 Krylov subspace methods

At step $k \geq 1$, a *Krylov subspace method* selects an approximation x_k of the solution x of (1.1) from the shifted Krylov space $x_0 + \mathcal{K}_k(y - Tx_0; T)$; here, x_0 is some initial guess of $T^\dagger y$, and the kth Krylov subspace $\mathcal{K}_k(z; T)$ is defined as the linear space

$$\mathcal{K}_k(z; T) = \mathrm{span}\{z, Tz, T^2 z, \ldots, T^{k-1} z\}.$$

Expanding $x_k - x_0$ in terms of the spanning elements of the Krylov subspace, the coefficients of a polynomial $q_{k-1} \in \Pi_{k-1}$ (the space of polynomials of degree $k - 1$; $\Pi_{-1} = \{0\}$) are found such that

$$x_k = x_0 + q_{k-1}(T)(y - Tx_0). \tag{2.1}$$

Associated with this sequence $\{q_{k-1}\}$ of *iteration polynomials* is another sequence of polynomials $\{p_k\}$ given by

$$p_k(\lambda) = 1 - \lambda q_{k-1}(\lambda).\qquad(2.2)$$

Obviously, $p_k \in \Pi_k^0$, where Π_k^0 denotes the set of normalized polynomials of degree k, i.e.,

$$\Pi_k^0 = \{p \in \Pi_k \,|\, p(0) = 1\}.$$

p_k is called *residual polynomial* since the residual $y - Tx_k$ satisfies

$$
\begin{aligned}
y - Tx_k &= y - T(x_0 + q_{k-1}(T)(y - Tx_0)) \\
&= y - Tx_0 - Tq_{k-1}(T)(y - Tx_0) \\
&= (I - Tq_{k-1}(T))(y - Tx_0) \\
&= p_k(T)(y - Tx_0).
\end{aligned}
$$

Moreover, if $y \in \mathcal{R}(T)$, i.e., if $y = Tx$ for some $x \in \mathcal{X}$ then

$$
\begin{aligned}
x - x_k &= x - x_0 - q_{k-1}(T)T(x - x_0) \\
&= p_k(T)(x - x_0).
\end{aligned}
$$

A key step to an efficient implementation of a Krylov subspace method lies in a cheap recursion for the computation of x_{k+1}, given $x_0, \ldots, x_k$. From this point of view, real orthogonal residual polynomials are especially convenient because they satisfy a three-term recurrence relation: if $k \geq 1$ and p_{k-1}, p_k and p_{k+1} are three consecutive orthogonal residual polynomials then, in view of their normalization at $\lambda = 0$, there are numbers $\alpha_k \neq 0$ and β_k, $k \geq 0$, such that

$$
\begin{aligned}
p_0 &= 1, \qquad p_1 = 1 - \alpha_0\lambda, \\
p_{k+1} &= -\alpha_k\lambda p_k + p_k - \alpha_k\frac{\beta_k}{\alpha_{k-1}}(p_{k-1} - p_k), \qquad k \geq 1,
\end{aligned}
\qquad(2.3)
$$

and hence the associated iteration polynomials satisfy

$$
\begin{aligned}
q_{-1} &= 0, \qquad q_0 = \alpha_0, \\
q_k &= q_{k-1} + \alpha_k(p_k + \frac{\beta_k}{\alpha_{k-1}}(q_{k-1} - q_{k-2})), \qquad k \geq 1.
\end{aligned}
$$

It follows from (2.1) that x_k can be computed by the following coupled recursion:

$$
\begin{aligned}
\Delta x_0 &= y - Tx_0, \\
x_1 &= x_0 + \alpha_0\Delta x_0, \\
\Delta x_k &= y - Tx_k + \beta_k\Delta x_{k-1}, \\
x_{k+1} &= x_k + \alpha_k\Delta x_k, \qquad k \geq 1.
\end{aligned}
\qquad(2.4)
$$

Consequently, only the two most recent iterates need to be stored, and the computation of x_{k+1} amounts to one application of the operator T, essentially.

8

One is left with the freedom of choosing the inner product $[\cdot,\cdot]$ for the orthogonality of $\{p_k\}$. Two conceptually different possibilities have to be distinguished. For instance, one can choose a fixed inner product a priori, thus giving rise to a fixed sequence of residual polynomials $\{p_k\}$, once and for all. This is the idea behind the so-called semiiterative methods, with the ν-methods being their most prominent representatives (cf. [6, 34]); here, the residual polynomials are rescaled and translated Jacobi polynomials over $[0, 1]$, and a lot of theoretical results are applicable for the analysis of these iterative schemes.

The disadvantage with the approach of using a prefixed weight function is the loss of adaptivity to special features of the right-hand side. For example, it may occur that the right-hand side has no components corresponding to larger spectral elements of T, that is, virtually, one is concerned with an operator whose spectrum is contained in a proper subinterval of $[0, 1]$. If such information were at hand, then there would be potential for speeding up the iterative process. Conjugate gradient type methods are adaptive in this sense, since the inner product $[\cdot,\cdot]$ is based on the spectral distribution of the right-hand side with respect to the spectral family $\{E_\lambda\}$ of T.

For $\varphi, \psi \in \Pi_\infty$ consider the family of bilinear forms

$$[\varphi, \psi]_n = \int_0^\infty \varphi(\lambda)\psi(\lambda)\, \lambda^n d\| E_\lambda(y - Tx_0)\|^2\,, \tag{2.5}$$

where the parameter n belongs to $\mathbf{N}_0$. Note the equivalent expression

$$[\varphi, \psi]_n = \langle \varphi(T)(y - Tx_0), T^n \psi(T)(y - Tx_0)\rangle$$

which is used for a numerical realization. It will prove convenient to extend the definition (2.5) to negative integers n by

$$[\varphi, \psi]_n = \int_{0+}^\infty \varphi(\lambda)\psi(\lambda)\, \lambda^n d\| E_\lambda(y - Tx_0)\|^2, \qquad n < 0\,,$$

which can formally be rewritten as

$$[\varphi, \psi]_n = \langle \varphi(T)(y - Tx_0), T^{\dagger|n|} \psi(T)(y - Tx_0)\rangle\,.$$

Of course, this definition for $n < 0$ is only well defined when $[1, 1]_n$ exists, i.e., when $y - Tx_0$ belongs to the domain of $(T^\dagger)^{|n|/2}$.

For those $n \in \mathbf{Z}$ for which $[1, 1]_n$ exists, e.g., for $n \in \mathbf{N}_0$, there is a well-defined sequence of orthogonal polynomials $\{p_k^{[n]}\}$ with $p_k^{[n]} \in \Pi_k^0$ and

$$[p_k^{[n]}, p_j^{[n]}]_n = 0, \qquad k \neq j\,. \tag{2.6}$$

In the sequel, a Krylov subspace method (2.1) will be called a *conjugate gradient type method*, if its residual polynomials are given by $\{p_k^{[n]}\}$ for some $n \in \mathbf{N}_0$. Later, in Chapter 6, it will be necessary to slightly modify this definition when studying problems with indefinite operators T. The superscript n will be omitted as long as there is no danger of confusion.

To implement a conjugate gradient method via (2.4), the coefficients α_k and β_k from (2.3) are required. As will be shown next, these parameters can be determined in the course of the iteration. It is obvious from (2.4) that

$$\Delta x_k = s_k(T)(y - Tx_0), \tag{2.7}$$

with

$$s_0 = 1, \qquad s_k = p_k + \beta_k s_{k-1}, \quad k \geq 1.$$

There is another relation between $\{s_k\}$ and $\{p_k\}$ which follows easily from (2.4), namely

$$p_{k+1} = p_k - \alpha_k \lambda s_k, \qquad k \geq 0. \tag{2.8}$$

Using the orthogonality of $\{p_k\}$, the previous two relations yield (for all $k \geq 0$)

$$0 = [p_{k+1}, s_k]_n = [p_k, s_k]_n - \alpha_k[\lambda s_k, s_k]_n = [p_k, p_k]_n - \alpha_k[s_k, s_k]_{n+1},$$

which enables the computation of α_k from

$$\alpha_k = \frac{[p_k, p_k]_n}{[s_k, s_k]_{n+1}}, \qquad k \geq 0. \tag{2.9}$$

It will be shown in Proposition 2.5 below that $\{s_k\}$ is again a sequence of orthogonal polynomials, namely one with respect to $[\cdot, \cdot]_{n+1}$. Hence, for $n \geq 1$,

$$\begin{aligned}
0 = [s_k, s_{k-1}]_{n+1} &= [p_k, \lambda s_{k-1}]_n + \beta_k[s_{k-1}, s_{k-1}]_{n+1} \\
&= \frac{1}{\alpha_{k-1}}[p_k, p_{k-1} - p_k]_n + \beta_k[s_{k-1}, s_{k-1}]_{n+1} \\
&= -\frac{1}{\alpha_{k-1}}[p_k, p_k]_n + \beta_k[s_{k-1}, s_{k-1}]_{n+1}.
\end{aligned}$$

From (2.9) therefore follows

$$\beta_k = \frac{1}{\alpha_{k-1}} \frac{[p_k, p_k]_n}{[s_{k-1}, s_{k-1}]_{n+1}} = \frac{[p_k, p_k]_n}{[p_{k-1}, p_{k-1}]_n}, \qquad k \geq 1. \tag{2.10}$$

Notice that β_0 is not required in (2.4).

As will become clear in the following chapters, the numbers $|p_k'(0)| = q_{k-1}(0)$ play an important role when conjugate gradient type methods are applied to regularize an ill-posed problem. Since $p_k(0) = 1$ and p_k has all k zeros in the convex hull of the spectrum of T which, by assumption, is contained in $[0, 1]$, it is obvious that $p_k'(0)$ is always negative for $k \geq 1$. Taking derivatives in (2.3) yields the recursion

$$|p_0'(0)| = 0, \qquad |p_1'(0)| = \alpha_0 \,,$$

$$|p_{k+1}'(0)| = \alpha_k + |p_k'(0)| + \alpha_k \frac{\beta_k}{\alpha_{k-1}} (|p_k'(0)| - |p_{k-1}'(0)|) \,, \qquad k \geq 1 \,. \tag{2.11}$$

Throughout this text let always be

$$\kappa = \text{number of } \textit{nonzero} \text{ points of increase of } \alpha(\lambda) \equiv \|E_\lambda(y - Tx_0)\|^2 \,. \tag{2.12}$$

κ may be finite of infinity, but for $\kappa < \infty$ the bilinear form $[\cdot, \cdot]_n$ fails to be definite on the space of all polynomials. It defines an inner product on $\Pi_{\kappa-1}$, though, and if the origin is no further point of increase, then there is a unique polynomial p_κ of degree κ in Π_κ^0 which is perpendicular to $\Pi_{\kappa-1}$. p_κ has all its κ roots in the points of increase of $\|E_\lambda(y - Tx_0)\|^2$, hence $[p_\kappa, p_\kappa]_n = 0$. In other words,

$$\|y - Tx_\kappa\| = \|p_\kappa(T)(y - Tx_0)\| = 0 \,,$$

independent of the parameter $n \in \mathbb{N}_0$, and conjugate gradient type methods terminate after κ steps with a solution x_κ of (1.1). Note that in this case $y - Tx_0$ belongs to an invariant subspace of T of dimension κ.

If y has a nontrivial component in the orthogonal complement of $\mathcal{R}(T)$ then $\lambda = 0$ is an additional (i.e., the $\kappa + 1$st) point of increase of $\|E_\lambda(y - Tx_0)\|$. In this case, the conclusions $y - Tx_\kappa = p_\kappa(T)y = E_0 y$, and $(I - E_0)x_\kappa = T^\dagger y$ remain valid when $n \geq 1$, since for $n \geq 1$ there is no contribution to $[\cdot, \cdot]_n$ coming from $\lambda = 0$. On the other hand, if $n = 0$ and the origin is the $\kappa + 1$st point of increase, then the roots of p_κ interlace with all points of increase, and no polynomial in $\Pi_{\kappa+1}^0$ is orthogonal to Π_κ, since any such polynomial should have a root at $\lambda = 0$. In other words, in this case $(I - E_0)x_\kappa \neq T^\dagger y$.

In any case, the iteration (2.4) "breaks down" in the course of the $\kappa + 1$st step since, by the above discussion, $[s_\kappa, s_\kappa]_{n+1}$ vanishes so that α_κ is undefined, cf. (2.9). Therefore $k = \kappa$ is the ultimate stopping index. This "irregular" finite termination will cause some extra considerations later on.

Conjugate gradient type methods combine in an ideal fashion computational simplicity (as shown above) with certain optimality properties. This is the essence of the following well-known fact.

Proposition 2.1 *Let x_k be the kth iterate ($0 \leq k \leq \kappa$) of the conjugate gradient type method with parameter $n \geq 1$, and let x be any other element from the same Krylov subspace $x_0 + \mathcal{K}_{k-1}(y - Tx_0; T)$. Then,*

$$\|T^{(n-1)/2}(y - Tx_k)\| \leq \|T^{(n-1)/2}(y - Tx)\| \,, \tag{2.13}$$

and equality holds if and only if $x = x_k$.

Proof. Rewriting (2.13) in terms of the residual polynomials p_k yields the equivalent statement

$$[p_k, p_k]_{n-1} \leq [p, p]_{n-1} \qquad \text{for all } p \in \Pi_k^0 . \tag{2.14}$$

Given any $p \in \Pi_k^0$ it follows that $p - p_k = \lambda s$ for some polynomial $s \in \Pi_{k-1}$, and hence, by orthogonality

$$[p, p]_{n-1} - [p_k, p_k]_{n-1} = [p - p_k, p + p_k]_{n-1} = [s, \lambda s + 2p_k]_n = [s, s]_{n+1} .$$

From this follows (2.13), and equality holds if and only if $s = 0$, i.e., if $p \equiv p_k$. $\qquad\square$

The polynomial $p \in \Pi_k^0$ minimizing (2.14) is the so-called *kernel polynomial* associated with the weight function $d\alpha = \lambda^{n-1} \, d\|E_\lambda(y - Tx_0)\|^2$, cf., e.g., [9, Sect. I.7] or [76, Sect. 3.1].

Remark. The case $n = 0$ plays an exceptional but interesting role: if $y \in \mathcal{R}(T^{1/2})$ then $\|T^{-1/2}(y - Tx_k)\|^2 = [p_k, p_k]_{-1}$ in (2.13) is well defined, and Proposition 2.1 and its proof remain valid. In particular, this is the case when (1.1) has a solution. In general, however, this assumption will not be satisfied. In this case a considerably weaker result holds, cf. Theorem 2.2.

2.2 Two particular conjugate gradient type methods

Although the properties of the different conjugate gradient type methods are very similar, the evaluation of the inner product (2.5) becomes somewhat expensive when n is large. Therefore, in practice, n is typically zero or one. The corresponding algorithms are discussed below in greater detail.

— The minimal residual method (MR)

Consider first the case $n = 1$. Here, the statement of Proposition 2.1 is especially attractive since the norms in (2.13) and (2.14) simplify:

$$[p, p]_0 = \|p(T)(y - Tx_0)\|^2 .$$

Consequently, Proposition 2.1 states that the residual $y - Tx_k$ of the kth iterate x_k has minimal norm among all iterates in the Krylov subspace $x_0 + \mathcal{K}_{k-1}(y - Tx_0; T)$. For this reason this method is occasionally called *minimal residual method* (MR).

Section 3.3 deals with a stopping rule for conjugate gradient type methods (with parameter $n \geq 1$), which halts the iteration as soon as the residual drops below a certain tolerance. Obviously, among all conjugate gradient type methods, MR will meet this stopping criterion first.

$$r_0 = y - Tx_0$$
$$d = r_0$$
$$Td = Tr_0$$
$$k = 0$$
while (not stop) do
$$\quad \alpha = \langle r_k, Tr_k \rangle / \|Td\|^2$$
$$\quad x_{k+1} = x_k + \alpha d$$
$$\quad r_{k+1} = r_k - \alpha Td$$
$$\quad \beta = \langle r_{k+1}, Tr_{k+1} \rangle / \langle r_k, Tr_k \rangle$$
$$\quad d = r_{k+1} + \beta d$$
$$\quad Td = Tr_{k+1} + \beta Td$$
$$\quad k = k + 1$$
end while.

Algorithm 2.1: MR

$$r_0 = y - Tx_0$$
$$d = r_0$$
$$k = 0$$
while (not stop) do
$$\quad \alpha = \|r_k\|^2 / \langle d, Td \rangle$$
$$\quad x_{k+1} = x_k + \alpha d$$
$$\quad r_{k+1} = r_k - \alpha Td$$
$$\quad \beta = \|r_{k+1}\|^2 / \|r_k\|^2$$
$$\quad d = r_{k+1} + \beta d$$
$$\quad k = k + 1$$
end while.

Algorithm 2.2: CG

The MR-method is summarized in Algorithm 2.1; by providing extra storage for Td only one multiplication with T is necessary per iteration.

— The conjugate gradient method (CG)

The conjugate gradient method (CG) as proposed originally by HESTENES and STIEFEL [45], cf. Algorithm 2.2, corresponds to the choice $n = 0$ in Section 2.1. When $y = Tx$, that is, when problem (1.1) is solvable, then the kth iterate x_k minimizes the error $x - x_k$ in the so-called *energy norm* $\langle x - x_k, T(x - x_k) \rangle$ in $x_0 + \mathcal{K}_{k-1}(y - Tx_0; T)$. When there is no solution (note that this is the typical situation in ill-posed problems if y is subject to noise) then one might expect that x_k somehow seeks to approximate "something unbounded". Nevertheless, CG can be applied successfully with an adequate stopping rule, cf. Section 4.3.

The following result is the analog of Proposition 2.1. It is included merely for the sake of completeness; it will not be used in the subsequent analysis. To state the theorem properly, more notation needs to be introduced: let $\{T_\varepsilon^\dagger\}_{\varepsilon>0}$ be the regularization corresponding to *truncated spectral expansion*, i.e.,

$$T_\varepsilon^\dagger y = x_0 + T^\dagger(I - E_\varepsilon)(y - Tx_0). \tag{2.15}$$

Theorem 2.2 *Let p_k be the kth residual polynomial of CG and x_k be the corresponding approximation. Furthermore, let x be any other element in $x_0 + \mathcal{K}_k(y - Tx_0; T)$ and let $p \in \Pi_k^0$ be the corresponding residual polynomial for which $y - Tx = p(T)(y - Tx_0)$. Then, with $T_\varepsilon^\dagger$ as in (2.15), the following holds for every $0 < \varepsilon \leq 1$:*

$$\|T^{1/2}(T_\varepsilon^\dagger y - x_k)\|^2 \leq \|T^{1/2}(T_\varepsilon^\dagger y - x)\|^2 + 2\int_0^\varepsilon \frac{(p - p_k)(\lambda)}{\lambda} d\|E_\lambda(y - Tx_0)\|^2. \tag{2.16}$$

Proof. Let $\varepsilon > 0$ be fixed and, for the sake of simplicity, let $d\alpha = d\|E_\lambda(y - Tx_0)\|^2$. Then,

$$
\begin{aligned}
\|T^{1/2}(T_\varepsilon^\dagger y - x)\|^2 &= \|E_\varepsilon T^{1/2}(T_\varepsilon^\dagger y - x)\|^2 + \|(I - E_\varepsilon)T^{1/2}(T_\varepsilon^\dagger y - x)\|^2 \\
&= \|E_\varepsilon T^{1/2}x\|^2 + \|(I - E_\varepsilon)T^{1/2}(T_\varepsilon^\dagger y - x)\|^2 \\
&= \int_0^\varepsilon \frac{(1 - p(\lambda))^2}{\lambda} d\alpha + \int_{\varepsilon+}^\infty \frac{p^2(\lambda)}{\lambda} d\alpha.
\end{aligned}
$$

Accordingly, when $p = p_k$, i.e., when $x = x_k$,

$$\|T^{1/2}(T_\varepsilon^\dagger y - x_k)\|^2 = \int_0^\varepsilon \frac{(1 - p_k(\lambda))^2}{\lambda} d\alpha + \int_{\varepsilon+}^\infty \frac{p_k^2(\lambda)}{\lambda} d\alpha. \tag{2.17}$$

14

By orthogonality (note that $(p_k - p)/\lambda \in \Pi_{k-1}$),

$$\int_0^\infty \frac{(p_k^2 - p^2)(\lambda)}{\lambda} \, d\alpha = [p_k + p, \frac{p_k - p}{\lambda}]_0 = [p - p_k, \frac{p_k - p}{\lambda}]_0 \leq 0 \,,$$

and consequently,

$$\int_{\varepsilon+}^\infty \frac{p_k^2(\lambda)}{\lambda} \, d\alpha \leq \int_{\varepsilon+}^\infty \frac{p^2(\lambda)}{\lambda} \, d\alpha + \int_0^\varepsilon \frac{(p^2 - p_k^2)(\lambda)}{\lambda} \, d\alpha \,.$$

Inserting this into (2.17) yields

$$\|T^{1/2}(T_\varepsilon^\dagger y - x_k)\|^2$$

$$\leq \int_0^\varepsilon \frac{(1 - p_k(\lambda))^2}{\lambda} \, d\alpha + \int_{\varepsilon+}^\infty \frac{p^2(\lambda)}{\lambda} \, d\alpha + \int_0^\varepsilon \frac{(p^2 - p_k^2)(\lambda)}{\lambda} \, d\alpha$$

$$= \|T^{1/2}(T_\varepsilon^\dagger y - x)\|^2 + \int_0^\varepsilon \left(\frac{(1 - p_k(\lambda))^2}{\lambda} + \frac{(p^2 - p_k^2)(\lambda)}{\lambda} - \frac{(1 - p(\lambda))^2}{\lambda} \right) d\alpha$$

$$= \|T^{1/2}(T_\varepsilon^\dagger y - x)\|^2 + 2 \int_0^\varepsilon \frac{(p - p_k)(\lambda)}{\lambda} \, d\alpha \,,$$

as was to be shown. □

Theorem 2.2 indicates that, to some extent, the iterates of CG approximate some $T_\varepsilon^\dagger y$ in the respective Krylov subspace. The number ε, however, has to be chosen carefully.

If $y \in \mathcal{R}(T)$, then one can let $\varepsilon \to 0$ in Theorem 2.2. In this case, the integral on the right-hand side of (2.16) goes to zero showing that Proposition 2.1 holds for $n = 0$ as well.

2.3 Conjugate gradient type methods using TT^*

So far it has been a basic assumption that T is selfadjoint and semidefinite. When this fails to be the case, conjugate gradient type methods lose all their properties and may break down with division by zero. However, as TT^* is selfadjoint and positive semidefinite, the previous algorithms may be applied to the (formal) problem

$$TT^* w = y, \qquad x = T^* w \,. \tag{2.18}$$

Denoting by $\{E_\lambda\}$ the spectral family of TT^*, the bilinear forms $[\cdot, \cdot]_n$ from (2.5) and the definition the corresponding residual polynomials for conjugate gradient

15

type algorithms immediately carry over to this more general situation. Obviously, $[\varphi, \psi]_n$ can be computed as

$$[\varphi, \psi]_n = \langle \varphi(TT^*)(y - Tx_0), (TT^*)^n \psi(TT^*)(y - Tx_0) \rangle,$$

where $(TT^*)^n$ is defined via the generalized inverse $(TT^*)^\dagger$ for $n < 0$.

The iterates of the resulting algorithms are defined as

$$w_k = w_0 + q_{k-1}(TT^*)(y - Tx_0),$$

where the meaning of w_0 remains to be clarified. Keeping in mind that the introduction of w was just a formal means to represent $x = T^*w$, the iteration should be transformed into x-space, which yields the (well-defined) recursion

$$x_k = x_0 + T^* q_{k-1}(TT^*)(y - Tx_0) = x_0 + q_{k-1}(T^*T)T^*(y - Tx_0).$$

x_k now belongs to a Krylov subspace with respect to T^*T, namely

$$x_k \in x_0 + \mathcal{K}_{k-1}(T^*(y - Tx_0); T^*T).$$

The residual $y - Tx_k$ fulfills

$$y - Tx_k = p_k(TT^*)(y - Tx_0),$$

and if $y = Tx$ then

$$x - x_k = p_k(T^*T)(x - x_0).$$

For $n \geq 1$ the iterate x_k only depends on the component of y in the closure of $\mathcal{R}(T)$ since $T^*y = T^*Py$; recall that P is the orthoprojector onto the closure of $\mathcal{R}(T)$. For $n = 0$ this is different because the bilinear form $[\cdot, \cdot]_0$ and thus the polynomials $\{p_k\}$ also depend on $(I - P)y$.

In the following, the most important representatives of this alternative class of conjugate gradient type algorithms are treated in greater detail.

— CG applied to the normal equation (CGNE)

When T fails to be selfadjoint, positive semidefinite, the standard approach is to apply CG to the *normal equation*

$$T^*Tx = T^*y.$$

The algorithm, cf. Algorithm 2.3, is called CGNE and corresponds to a conjugate gradient type method (for TT^*) as introduced above, namely the one for $n = 1$. According to Proposition 2.1, the residual polynomials minimize $[p_n, p_n]_0$; hence, the iterates x_k minimize the residual

$$\|y - Tx_k\| = \|p_k(TT^*)(y - Tx_0)\|.$$

Consequently, CGNE shares the following property with MR: if the iteration is to be terminated as soon as the residual drops below a given tolerance, then CGNE will take fewest iterations among all conjugate gradient type methods of the above form.

$$r_0 = y - Tx_0$$
$$d = T^*r_0$$
$$k = 0$$
while (not stop) do
$$\alpha = \|T^*r_k\|^2 / \|Td\|^2$$
$$x_{k+1} = x_k + \alpha d$$
$$r_{k+1} = r_k - \alpha Td$$
$$\beta = \|T^*r_{k+1}\|^2 / \|T^*r_k\|^2$$
$$d = T^*r_{k+1} + \beta d$$
$$k = k + 1$$
end while.

Algorithm 2.3: CGNE

$$r_0 = y - Tx_0$$
$$d = T^*r_0$$
$$k = 0$$
while (not stop) do
$$\alpha = \|r_k\|^2 / \|d\|^2$$
$$x_{k+1} = x_k + \alpha d$$
$$r_{k+1} = r_k - \alpha Td$$
$$\beta = \|r_{k+1}\|^2 / \|r_k\|^2$$
$$d = T^*r_{k+1} + \beta d$$
$$k = k + 1$$
end while.

Algorithm 2.4: CGME

— The minimal error method (CGME)

CG applied straight to (2.18) corresponds to the choice $n = 0$ (cf. Algorithm 2.4). In general, the iterates have no optimality property but for $y \in \mathcal{R}(T)$ they minimize the error norm $\|T^{\dagger}y - x_k\|$ in the respective Krylov space. This algorithm shall therefore be called *minimal error method* (CGME). For perturbed data, the analog of Theorem 2.2 reads as follows:

Theorem 2.3 *Let p_k be the kth residual polynomial and x_k the kth approximation of CGME. Furthermore, let x be any other element in $x_0 + \mathcal{K}_k(T^*(y - Tx_0); T^*T)$ and let $p \in \Pi_k^0$ be the corresponding residual polynomial. Then, with $T_\varepsilon^{\dagger}$ as in (2.15), the following holds for every $0 < \varepsilon \leq 1$:*

$$\|T_\varepsilon^{\dagger}y - x_k\|^2 \leq \|T_\varepsilon^{\dagger}y - x\|^2 + 2\int_0^\varepsilon \frac{(p - p_k)(\lambda)}{\lambda} \, d\|E_\lambda(y - Tx_0)\|^2 \, .$$

The proof is completely analogous to the proof of Theorem 2.2 and is left to the reader.

2.4 Basic relations between conjugate gradient type methods

In the remainder, the analysis will be restricted to the case $x_0 = 0$. Since the general case of a nonvanishing initial guess can be transformed to this special case via the transformation $x \mapsto x - x_0$, $y \mapsto y - Tx_0$, there is no loss of generality. Furthermore, only the results for the algorithms of Section 2.1 for selfadjoint, semidefinite operators T shall be stated explicitly. If no opposite is said, the results hold accordingly for the algorithms of Section 2.3 for non-selfadjoint T, and the proofs apply almost word by word. Significant differences will be mentioned.

There are a number of quite interesting connections between the residual polynomials of conjugate gradient type methods corresponding to different parameters n. To state these results properly, recall the notation $\{p_k^{[n]}\}$ of (2.6) for the residual polynomials that are orthogonal with respect to $[\cdot, \cdot]_n$. From the theory of real orthogonal polynomials it is well known that $p_k^{[n]}$ has k simple real zeros $\lambda_{j,k}^{[n]}$, $j = 1, \ldots, k$, with

$$0 < \lambda_{1,k}^{[n]} < \lambda_{2,k}^{[n]} < \ldots < \lambda_{k,k}^{[n]} \leq \|T\| \leq 1 \, .$$

The superscript parameter n will again be omitted when its value is clear from the context. It follows that

$$p_k^{[n]}(\lambda) = \prod_{j=1}^k (1 - \frac{\lambda}{\lambda_{j,k}^{[n]}}), \qquad p_k^{[n]\prime}(0) = -\sum_{j=1}^k \frac{1}{\lambda_{j,k}^{[n]}} \, . \tag{2.19}$$

18

Note that $p_k^{[n]'}(0)$ is always negative and $|p_k^{[n]'}(0)| \geq k$. Recall that the zeros of two consecutive polynomials interlace, i.e., one has

$$0 < \lambda_{1,k+1}^{[n]} < \lambda_{1,k}^{[n]} < \lambda_{2,k+1}^{[n]} < \lambda_{2,k}^{[n]} < \ldots < \lambda_{k,k}^{[n]} < \lambda_{k+1,k+1}^{[n]} \,.$$

The following statement is essentially the translation of a well-known identity for kernel polynomials from the orthogonal polynomial literature (cf. [76, Theorem 3.1.3]).

Lemma 2.4 *Let $n \in \mathbb{N}$, $0 \leq k < \kappa$. Then one has*

$$p_k^{[n]} = [p_k^{[n]}, p_k^{[n]}]_{n-1} \sum_{j=0}^{k} [p_j^{[n-1]}, p_j^{[n-1]}]_{n-1}^{-1} p_j^{[n-1]}, \tag{2.20}$$

and the minimum of (2.13) satisfies the identity

$$\|T^{(n-1)/2}(y - Tx_k)\|^2 = [p_k^{[n]}, p_k^{[n]}]_{n-1} = \left(\sum_{j=0}^{k} [p_j^{[n-1]}, p_j^{[n-1]}]_{n-1}^{-1}\right)^{-1}. \tag{2.21}$$

Proof. The first equality in (2.21) is clear from the definition of x_k. To verify the remaining assertions, expand $p_k^{[n]}$ in terms of $p_j^{[n-1]}$, which yields

$$p_k^{[n]} = \sum_{j=0}^{k} \frac{[p_k^{[n]}, p_j^{[n-1]}]_{n-1}}{[p_j^{[n-1]}, p_j^{[n-1]}]_{n-1}} p_j^{[n-1]} \,.$$

Since $p_j^{[n-1]}(0) = p_k^{[n]}(0) = 1$ one has $(p_j^{[n-1]} - p_k^{[n]})/\lambda \in \Pi_{k-1}$ for any j with $0 \leq j \leq k$, and hence, using the orthogonality of $\{p_k^{[n]}\}$ one obtains

$$[p_k^{[n]}, p_j^{[n-1]}]_{n-1} = [p_k^{[n]}, p_j^{[n-1]} - p_k^{[n]}]_{n-1} + [p_k^{[n]}, p_k^{[n]}]_{n-1}$$

$$= [p_k^{[n]}, \frac{p_j^{[n-1]} - p_k^{[n]}}{\lambda}]_n + [p_k^{[n]}, p_k^{[n]}]_{n-1} = [p_k^{[n]}, p_k^{[n]}]_{n-1} \,.$$

Inserting this into the expansion for $p_k^{[n]}$ gives (2.20). The second equality in (2.21) follows immediately from (2.20) evaluated at the origin, since $p_k^{[n]}(0) = p_j^{[n-1]}(0) = 1$ for all $0 \leq j \leq k$. $\qquad \square$

Remark. A closer look at the above proof shows that Lemma 2.4 extends to nonpositive integers n for which $[1,1]_n$ exists whenever

$$[p_k^{[n]}, p_j^{[n-1]} - p_k^{[n]}]_{n-1} = [p_k^{[n]}, \frac{p_j^{[n-1]} - p_k^{[n]}}{\lambda}]_n \,.$$

This is always the case for $n < 0$; it is true for $n = 0$, if and only if $E_0 y = 0$.

The next proposition reveals another connection between the two consecutive sequences $\{p_k^{[n]}\}$ and $\{p_k^{[n+1]}\}$. This result implies in particular that the polynomial s_k given in (2.7) and (2.8) is a scalar multiple of $p_k^{[n+1]}$.

Proposition 2.5 *Let* $n \in \mathbb{N}_0$, $0 \leq k < \kappa$. *Then*

$$p_k^{[n+1]} = \frac{1}{\pi_{k,n}} \frac{p_k^{[n]} - p_{k+1}^{[n]}}{\lambda} \quad with \quad \pi_{k,n} = p_k^{[n]}{}'(0) - p_{k+1}^{[n]}{}'(0) > 0. \tag{2.22}$$

Proof. By construction the right-hand side of (2.22) belongs to Π_k^0 ($\pi_{k,n}$ is positive because of (2.19) and the interlacing properties of the zeros of $p_k^{[n]}$ and $p_{k+1}^{[n]}$). Denote this polynomial by p. For any other polynomial $q \in \Pi_{k-1}$ one has

$$[p,q]_{n+1} = \frac{1}{\pi_{k,n}} [p_k^{[n]} - p_{k+1}^{[n]}, q]_n = \frac{1}{\pi_{k,n}} ([p_k^{[n]}, q]_n - [p_{k+1}^{[n]}, q]_n) = 0.$$

This proves that p equals $p_k^{[n+1]}$. $\qquad\square$

Lemma 2.4 and Proposition 2.5 - when rewritten in terms of *orthonormal* polynomials - are known as *Christoffel-Darboux identity* in the orthogonal polynomial literature. Later on it will become necessary to estimate $|p_k^{[n]}{}'(0)|$. To this end, the following corollary provides an alternative identity for $\pi_{k,n}$, which follows readily from Proposition 2.5. Note that this result requires $n \geq 1$.

Corollary 2.6 *For* $n \in \mathbb{N}$, $0 \leq k < \kappa$, *the following holds:*

$$\pi_{k,n} = p_k^{[n]}{}'(0) - p_{k+1}^{[n]}{}'(0) = \frac{[p_k^{[n]}, p_k^{[n]}]_{n-1} - [p_{k+1}^{[n]}, p_{k+1}^{[n]}]_{n-1}}{[p_k^{[n+1]}, p_k^{[n+1]}]_n}.$$

Proof. For $0 \leq k < \kappa$ and $n \geq 1$,

$$[p_k^{[n]}, 1]_{n-1} = [p_k^{[n]}, \frac{1 - p_k^{[n]}}{\lambda}]_n + [p_k^{[n]}, p_k^{[n]}]_{n-1} = [p_k^{[n]}, p_k^{[n]}]_{n-1}, \tag{2.23}$$

since $(1 - p_k^{[n]})/\lambda \in \Pi_{k-1}$. Thus, Proposition 2.5 yields

$$\begin{aligned}
[p_k^{[n+1]}, p_k^{[n+1]}]_n &= [p_k^{[n+1]}, 1]_n \\
&= ([p_k^{[n]}, 1]_{n-1} - [p_{k+1}^{[n]}, 1]_{n-1})/\pi_{k,n} \\
&= ([p_k^{[n]}, p_k^{[n]}]_{n-1} - [p_{k+1}^{[n]}, p_{k+1}^{[n]}]_{n-1})/\pi_{k,n},
\end{aligned}$$

and the statement of the corollary follows. $\qquad\square$

20

Another consequence of Proposition 2.5 is the following well-known relation between the zeros of kernel polynomials and orthogonal polynomials.

Corollary 2.7 *For* $1 \leq k < \kappa$ *and* $n \in \mathbf{N}_0$ *the following interlacing properties hold for the zeros of the polynomials* $p_k^{[n]}$, $p_{k+1}^{[n]}$, *and* $p_k^{[n+1]}$:

$$0 < \lambda_{1,k+1}^{[n]} < \lambda_{1,k}^{[n]} < \lambda_{1,k}^{[n+1]} < \lambda_{2,k+1}^{[n]} < \lambda_{2,k}^{[n]} < \lambda_{2,k}^{[n+1]} < \ldots < \lambda_{k,k}^{[n]} < \lambda_{k,k}^{[n+1]} < \lambda_{k+1,k+1}^{[n]} .$$

Proof. From (2.22) and the interlacing property of the zeros of $p_k^{[n]}$ and $p_{k+1}^{[n]}$ follows that $p_k^{[n+1]}$ has alternating signs at the zeros of $p_k^{[n]}$. Moreover, $p_k^{[n+1]}(\lambda_{1,k}^{[n]}) > 0$ must hold since $p_{k+1}^{[n]}(\lambda_{1,k}^{[n]}) < 0$. Similarly, $p_k^{[n+1]}$ has alternating signs at the zeros of $p_{k+1}^{[n]}$, which yields the desired inclusion for the zeros of $p_k^{[n+1]}$. □

There is also a result connecting *three* consecutive sequences of residual polynomials $\{p_k^{[n]}\}$, $\{p_k^{[n+1]}\}$ and $\{p_k^{[n+2]}\}$:

Proposition 2.8 *Let* $n \in \mathbf{N}_0$, $0 \leq k < \kappa - 1$. *Then*

$$p_k^{[n+2]} = \frac{1}{\theta_{k+1,n}} \frac{p_{k+1}^{[n+1]} - p_{k+1}^{[n]}}{\lambda} \quad with \quad \theta_{k+1,n} = p_{k+1}^{[n+1]\prime}(0) - p_{k+1}^{[n]\prime}(0) > 0. \quad (2.24)$$

Proof. Let $p := (p_{k+1}^{[n+1]} - p_{k+1}^{[n]})/\lambda$. It follows that $p \in \Pi_k$ with $p \not\equiv 0$ as long as $k < \kappa - 1$. Furthermore, for any polynomial $q \in \Pi_{k-1}$,

$$[p, q]_{n+2} = [p_{k+1}^{[n+1]}, q]_{n+1} - [p_{k+1}^{[n]}, \lambda q]_n = 0$$

by orthogonality (note that $\lambda q \in \Pi_k$). Hence, p is a scalar multiple of $p_k^{[n+2]}$; the representation of $\theta_{k+1,n}$ can be determined by letting $\lambda \to 0$. $\theta_{k+1,n}$ is positive because of (2.19) and the interlacing properties of the zeros of $p_{k+1}^{[n+1]}$ and $p_{k+1}^{[n]}$, cf. Corollary 2.7. □

Remark. If $k = \kappa - 1$ then three cases must be distinguished (recall the discussion in Section 2.1): if $n \geq 1$ then $p_\kappa^{[n+1]} = p_\kappa^{[n]}$, and hence (2.24) is no longer true for $k = \kappa - 1$ since the polynomial p constructed in the proof vanishes identically; the same argument applies when $n = 0$ and $y \in \mathcal{R}(T)$. If $n = 0$ and $y \notin \mathcal{R}(T)$ then $p_\kappa^{[1]} \neq p_\kappa^{[0]}$ so that $p \not\equiv 0$. In this case Proposition 2.8 remains true for $k = \kappa - 1$.

As in Corollary 2.6 one can use this result to derive an alternative expression for $\theta_{k,n}$. This identity will later on play a central role in the analysis of the CG method.

Corollary 2.9 *For $n \in \mathbb{N}_0$, $0 < k \leq \kappa$, the following holds:*

$$\theta_{k,n} = p_k^{[n+1]\prime}(0) - p_k^{[n]\prime}(0) = \frac{[p_k^{[n+1]}, p_k^{[n+1]}]_n}{[p_{k-1}^{[n+2]}, p_{k-1}^{[n+2]}]_{n+1}} \,. \tag{2.25}$$

Proof. Making use of (2.23), Proposition 2.8 yields for $0 < k < \kappa$:

$$\begin{aligned}
[p_{k-1}^{[n+2]}, p_{k-1}^{[n+2]}]_{n+1} &= [p_{k-1}^{[n+2]}, 1]_{n+1} \\
&= ([p_k^{[n+1]}, 1]_n - [p_k^{[n]}, 1]_n)/\theta_{k,n} \\
&= [p_k^{[n+1]}, 1]_n/\theta_{k,n} \\
&= [p_k^{[n+1]}, p_k^{[n+1]}]_n/\theta_{k,n} \,.
\end{aligned}$$

Therefore (2.25) is true for $0 < k < \kappa$. For $k = \kappa$, the above derivation remains valid if Proposition 2.8 can be applied. According to the remark following Proposition 2.8 this is the case if $p_\kappa^{[n+1]} \neq p_\kappa^{[n]}$. Otherwise both sides of (2.25) vanish, so that the statement is trivially fulfilled. □

2.5 Implementing both MR and CG in one scheme

The strong connections between the sequences $\{p_k^{[n]}\}$ and $\{p_k^{[n+1]}\}$ suggest the possibility of implementing the conjugate gradient methods with parameters n and $n + 1$ together in *one* scheme without doubling the number of multiplications with the operator T. In view of the different optimality properties of the two sequences this may provide a possibility to double check the quality of the computed approximations.

In this section it will be shown that it is indeed possible to extend the conjugate gradient type algorithms in this way. For ease of notation, however, it will only be shown how to implement MR and CG in one scheme. The extension to other values of n is straightforward. Throughout this section, x_k always denotes the kth iterate of CG, whereas the corresponding MR iterate is called z_k. The associated iteration polynomials inherit the superscript notation of the residual polynomials, i.e.,

$$x_k = q_{k-1}^{[0]}(T)y, \qquad z_k = q_{k-1}^{[1]}(T)y \,.$$

As it turns out, one can extend both algorithms, MR and CG so as to compute the other sequence of iterates as well.

Consider the MR scheme first. From Proposition 2.5 follows

$$q_k^{[0]} = \frac{1 - p_{k+1}^{[0]}}{\lambda} = q_{k-1}^{[0]} + \frac{p_k^{[0]} - p_{k+1}^{[0]}}{\lambda} = q_{k-1}^{[0]} + \pi_{k,0} p_k^{[1]},$$

and hence,

$$x_{k+1} = q_k^{[0]}(T)y = q_{k-1}^{[0]}(T)y + \pi_{k,0}p_k^{[1]}(T)y = x_k + \pi_{k,0}(y - Tz_k).$$

Note that the expression for $\pi_{k,0}$ given in Corollary 2.6 is not implementable within the MR scheme because it requires knowledge of the residual $p_{k+1}^{[0]}(T)y = y - Tx_{k+1}$. However, there is an alternative expression for $\pi_{k,0}$: by Proposition 2.5 one has

$$p_{k+1}^{[0]} = p_k^{[0]} - \pi_{k,0}\lambda p_k^{[1]},$$

and the orthogonality relations imply

$$0 = [p_{k+1}^{[0]}, p_k^{[1]}]_0 = [p_k^{[0]}, p_k^{[1]}]_0 - \pi_{k,0}[p_k^{[1]}, p_k^{[1]}]_1,$$

or equivalently,

$$\pi_{k,0} = \frac{[p_k^{[0]}, p_k^{[1]}]_0}{[p_k^{[1]}, p_k^{[1]}]_1}.$$

This can be further rewritten - similar to (2.23) - since

$$[p_k^{[0]}, p_k^{[1]}]_0 = [p_k^{[1]}, p_k^{[1]}]_0 + [\frac{p_k^{[0]} - p_k^{[1]}}{\lambda}, p_k^{[1]}]_1 = [p_k^{[1]}, p_k^{[1]}]_0.$$

In conclusion this yields the following recursion for the CG iterates within the MR scheme:

$$x_{k+1} = x_k + \frac{\|y - Tz_k\|^2}{\langle y - Tz_k, T(y - Tz_k)\rangle}(y - Tz_k).$$

No further extra work is required since all quantities in this recursion are computed anyway, cf. Algorithm 2.5.

An alternative implementation could start with Algorithm 2.2, i.e., with the CG scheme. Here, the key formula turns out to be (2.20) in Lemma 2.4. Note that the right-hand side of (2.20) is a convex combination of the residual polynomials $p_j^{[n-1]}$, $0 \leq j \leq k$. Therefore, it follows that

$$q_k^{[1]} = [p_{k+1}^{[1]}, p_{k+1}^{[1]}]_0 \sum_{j=0}^{k+1} [p_j^{[0]}, p_j^{[0]}]_0^{-1} q_{j-1}^{[0]} = \frac{[p_{k+1}^{[1]}, p_{k+1}^{[1]}]_0}{[p_k^{[1]}, p_k^{[1]}]_0} q_{k-1}^{[1]} + \frac{[p_{k+1}^{[1]}, p_{k+1}^{[1]}]_0}{[p_{k+1}^{[0]}, p_{k+1}^{[0]}]_0} q_k^{[0]},$$

and hence,

$$\begin{aligned} z_{k+1} &= q_k^{[1]}(T)y = \frac{[p_{k+1}^{[1]}, p_{k+1}^{[1]}]_0}{[p_k^{[1]}, p_k^{[1]}]_0} q_{k-1}^{[1]}(T)y + \frac{[p_{k+1}^{[1]}, p_{k+1}^{[1]}]_0}{[p_{k+1}^{[0]}, p_{k+1}^{[0]}]_0} q_k^{[0]}(T)y \\ &= \frac{[p_{k+1}^{[1]}, p_{k+1}^{[1]}]_0}{[p_k^{[1]}, p_k^{[1]}]_0} z_k + \frac{[p_{k+1}^{[1]}, p_{k+1}^{[1]}]_0}{[p_{k+1}^{[0]}, p_{k+1}^{[0]}]_0} x_{k+1}. \end{aligned}$$

23

$$x_0 = z_0$$
$$r_0 = y - T z_0$$
$$d = r_0$$
$$T d = T r_0$$
$$k = 0$$
while (not stop) do
$$\alpha = \langle r_k, T r_k \rangle / \| T d \|^2$$
$$z_{k+1} = z_k + \alpha d$$
$$\pi = \| r_k \|^2 / \langle r_k, T r_k \rangle$$
$$x_{k+1} = x_k + \pi r_k$$
$$r_{k+1} = r_k - \alpha T d$$
$$\beta = \langle r_{k+1}, T r_{k+1} \rangle / \langle r_k, T r_k \rangle$$
$$d = r_{k+1} + \beta d$$
$$T d = T r_{k+1} + \beta T d$$
$$k = k + 1$$
end while.

Algorithm 2.5: MR + CG

$$z_0 = x_0$$
$$r_0 = y - T x_0$$
$$d = r_0$$
$$\gamma_0 = 1$$
$$k = 0$$
while (not stop) do
$$\alpha = \| r_k \|^2 / \langle d, T d \rangle$$
$$x_{k+1} = x_k + \alpha d$$
$$r_{k+1} = r_k - \alpha T d$$
$$\beta = \| r_{k+1} \|^2 / \| r_k \|^2$$
$$\gamma_{k+1} = 1 + \beta \gamma_k$$
$$z_{k+1} = (\beta \gamma_k z_k + x_{k+1}) / \gamma_{k+1}$$
$$d = r_{k+1} + \beta d$$
$$k = k + 1$$
end while.

Algorithm 2.6: CG + MR

In this expression the numerator $[p_{k+1}^{[1]}, p_{k+1}^{[1]}]_0$ can not be evaluated as $p_{k+1}^{[1]}$ is not yet known. This can be remedied by another application of (2.21) in Lemma 2.4. Introducing

$$\gamma_k = \frac{[p_k^{[0]}, p_k^{[0]}]_0}{[p_k^{[1]}, p_k^{[1]}]_0}, \tag{2.26}$$

the factor in front of z_k can be rewritten as

$$\frac{[p_{k+1}^{[1]}, p_{k+1}^{[1]}]_0}{[p_k^{[1]}, p_k^{[1]}]_0} = \frac{\gamma_k \beta_{k+1}}{\gamma_{k+1}}$$

with the same β_k's as in (2.10) (note that $n = 0$ in the present context). Because of (2.21) the sequence $\{\gamma_k\}$ enjoys the following recursion for $k \geq 0$,

$$\begin{aligned}
\gamma_{k+1} &= [p_{k+1}^{[0]}, p_{k+1}^{[0]}]_0 \sum_{j=0}^{k+1} [p_j^{[0]}, p_j^{[0]}]_0^{-1} = 1 + [p_{k+1}^{[0]}, p_{k+1}^{[0]}]_0 \sum_{j=0}^{k} [p_j^{[0]}, p_j^{[0]}]_0^{-1} \\
&= 1 + \frac{[p_{k+1}^{[0]}, p_{k+1}^{[0]}]_0}{[p_k^{[1]}, p_k^{[1]}]_0} = 1 + \frac{[p_{k+1}^{[0]}, p_{k+1}^{[0]}]_0}{[p_k^{[0]}, p_k^{[0]}]_0} \gamma_k = 1 + \beta_{k+1} \gamma_k.
\end{aligned}$$

Thus, the MR iterates can be computed within the CG scheme as follows, cf. Algorithm 2.6:

$$\gamma_0 = 1, \quad \gamma_{k+1} = 1 + \beta_{k+1} \gamma_k, \quad z_{k+1} = \beta_{k+1} \frac{\gamma_k}{\gamma_{k+1}} z_k + \frac{1}{\gamma_{k+1}} x_{k+1}.$$

It should be remarked that Algorithm 2.6 requires one inner-product less than Algorithm 2.5, and also no update for Td; it is as cheap as the "pure" MR implementation in Algorithm 2.1. For the stopping rules suggested in the later chapters the norm of the MR residuals are required. They are not computed in Algorithm 2.6 but they are immediately available since

$$\|y - Tz_k\|^2 = \frac{1}{\gamma_k} \|y - Tx_k\|^2,$$

cf. (2.26). As $\|y - Tx_k\|$ is known this requires no additional inner product.

It is also easy to incorporate the computation of both sequences $\{p_k^{[0]'}(0)\}$ and $\{p_k^{[1]'}(0)\}$. Since $|p_k^{[n]'}(0)| = q_{k-1}^{[n]}(0)$ these numbers can be derived from the recursion for the iterates. For example, in Algorithm 2.6 one would compute

$$|p_{k+1}^{[1]'}(0)| = (\beta_{k+1} \gamma_k |p_k^{[1]'}(0)| + |p_{k+1}^{[0]'}(0)|)/\gamma_{k+1}.$$

The conjugate gradient type methods of Section 2.3 can also be implemented within one scheme. For CGME and CGNE this is exemplified in Algorithms 2.7 and 2.8. As can be seen, Algorithm 2.8 requires one inner product less than Algorithm 2.7, but it needs four vector updates as opposed to three vector updates in CGNE (Algorithm 2.3).

$$x_0 = z_0$$
$$r_0 = y - Tz_0$$
$$d = T^*r_0$$
$$k = 0$$
while (not stop) do
$$\alpha = \|T^*r_k\|^2/\|Td\|^2$$
$$z_{k+1} = z_k + \alpha d$$
$$\pi = \|r_k\|^2/\|T^*r_k\|^2$$
$$x_{k+1} = x_k + \pi T^*r_k$$
$$r_{k+1} = r_k - \alpha Td$$
$$\beta = \|T^*r_{k+1}\|^2/\|T^*r_k\|^2$$
$$d = T^*r_{k+1} + \beta d$$
$$k = k + 1$$
end while.

Algorithm 2.7: CGNE + CGME

$$z_0 = x_0$$
$$r_0 = y - Tx_0$$
$$d = T^*r_0$$
$$\gamma_0 = 1$$
$$k = 0$$
while (not stop) do
$$\alpha = \|r_k\|^2/\|d\|^2$$
$$x_{k+1} = x_k + \alpha d$$
$$r_{k+1} = r_k - \alpha Td$$
$$\beta = \|r_{k+1}\|^2/\|r_k\|^2$$
$$\gamma_{k+1} = 1 + \beta\gamma_k$$
$$z_{k+1} = (\beta\gamma_k z_k + x_{k+1})/\gamma_{k+1}$$
$$d = T^*r_{k+1} + \beta d$$
$$k = k + 1$$
end while.

Algorithm 2.8: CGME + CGNE

2.6 Stability issues

There are a number of mathematically equivalent implementations of the methods of the former sections. For instance, two different implementations of MR have been derived in Section 2.2 and 2.5, i.e., Algorithms 2.1 and 2.6. Another implementation of MR that can be found in the literature differs in the definition of α_k for (2.8): since $p_{k+1} = p_k - \alpha_k \lambda s_k$ minimizes $[p_{k+1}, p_{k+1}]_0$ over Π_{k+1}^0, α_k may be determined from the condition

$$\frac{d}{d\alpha} [p_k - \alpha \lambda s_k, p_k - \alpha \lambda s_k]_0 = 0,$$

which gives

$$\alpha_k = \frac{[p_k, s_k]_1}{[s_k, s_k]_2}. \tag{2.27}$$

Finally, some authors advocate Lanczos based implementations like the SYMMLQ (or MINRES) algorithm of PAIGE and SAUNDERS [64]. Similar implementations exist for the methods of Section 2.3: LSQR [65], for example, is an equivalent implementation of CGNE based on Lanczos bidiagonalization.

The principal problem with any of these different implementations of one and the same method is the loss of orthogonality in the residuals (or residual polynomials) due to finite precision arithmetic. Recall that this orthogonality is the major tool for the analytical results in this book. However, as it turns out, orthogonality can only be maintained by reorthogonalization techniques that are significantly more expensive and require a larger number of intermediate vectors (cf., e.g., [26, Sect. 9.2]). In the literature the influence of round-off errors on conjugate gradient type methods has been studied mainly for well-posed problems; only HANSEN [42] comments on the ill-posed case. This distinction is important though, because ill-posed problems obviously require completely different standards.

In the numerical experiments for this book CGNE and its variant with α_k computed by (2.27), Algorithm 2.8, and LSQR have been compared with each other. As it turns out, up to the point of divergence the error norms $\|x - x_k\|$ of the four algorithms always differed by less than about 5%, i.e., their difference is negligible as compared to the limited accuracy that can be achieved for ill-posed problems. Reorthogonalization has only been possible for the smaller problem in Section 6.7: the result was a slight speedup due to the lack of so-called "spurious eigenvalues", but the optimal accuracy was never better than without reorthogonalization. This is in agreement with the observations in [42]. In view of these limited experimental results it does not seem to pay to prefer the slightly more expensive implementation LSQR over Algorithm 2.3 for ill-posed problems, and Algorithm 2.8 seems to be as reliable.

Besides the question of numerical stability concerning round-off it is as important to study the influence of perturbations in the right-hand side. In the remainder of this section this will be considered in terms of continuity of the operator R_k that maps the right-hand side onto the corresponding kth iterate.

Proposition 2.10 *Let the selfadjoint semidefinite operator T be compact and non-degenerate, and consider any conjugate gradient type method with parameter $n \in \mathbf{N}_0$. Then, for any $k \in \mathbf{N}$, the operator R_k (that maps the right-hand side y onto the kth iterate x_k) is discontinuous in $\mathcal{X}$.*

Proof. Let $\{v_j\}_{j \in \mathbf{N}}$ be an orthonormal eigensystem for T, and let $\{\lambda_j\}$ be the corresponding eigenvalues, i.e., $Tv_j = \lambda_j v_j$. Without loss of generality let the first $k-1$ eigenvalues be nonzero and pairwise distinct. For

$$x = \sum_{j=1}^{k-1} \xi_j v_j, \qquad y = Tx,$$

with $\xi_j \neq 0$, $1 \leq j < k$, let $m \geq k$ be such that $\lambda_m \neq \lambda_j$, $1 \leq j < k$, and define

$$y^\pm = y \pm \delta v_m, \qquad \delta > 0.$$

As T is non-degenerate the above construction yields $y, y^\pm \in \mathcal{R}(T)$, and obviously $y^\pm \to y$ as $\delta \to 0$.

Denote by $\{x_k^\pm\}$ the iterates of the conjugate gradient type method corresponding to right-hand sides $y^\pm$, respectively. Accordingly, let $\{p_k^\pm\}$ denote the corresponding residual polynomials. Since $\|E_\lambda y^\pm\|^2$ has precisely k points of increase, none of which is at $\lambda = 0$, it follows from the discussion in Section 2.1 that

$$p_k^\pm(\lambda) = (1 - \frac{\lambda}{\lambda_m}) \prod_{j=1}^{k-1} (1 - \frac{\lambda}{\lambda_j}),$$

hence,

$$x_k^\pm = R_k y^\pm = T^\dagger y^\pm = x \pm \frac{\delta}{\lambda_m} v_m.$$

Choosing $m = m(\delta)$ such that $\delta/\lambda_{m(\delta)}$ diverges to infinity as $\delta \to 0$, this yields

$$\|R_k y^+ - R_k y^-\| = 2\delta/\lambda_{m(\delta)} \longrightarrow \infty, \qquad \delta \to 0,$$

proving thus the discontinuity of R_k. $\qquad\square$

Every stopping rule for a conjugate gradient type method must ultimately take care of the phenomenon discovered in the above proof. In particular, no a priori choice for k, i.e., a stopping rule depending only on δ, can render a conjugate gradient type method a regularization method. Since this (in)stability is such a delicate matter the following theorem provides a complete characterization of all points of discontinuity of R_k, valid for general (not necessarily compact) operators T.

Theorem 2.11 *Assume that (1.1) is ill-posed, i.e., that $T^\dagger$ is unbounded. Then the operator R_k, $k \in \mathbb{N}$, is discontinuous at y, if and only if Py belongs to a $(k-1)$-dimensional invariant subspace of T.*

Proof. Consider the *only-if-part* first, i.e., assume that Py does not belong to a $(k-1)$-dimensional invariant subspace of T, and let $\{y^\delta\}_{\delta > 0}$ be approximations of y with $y^\delta \to y^0 := y$ as $\delta \to 0$. Further on let the kth residual polynomial p_k^δ corresponding to the right-hand side y^δ, $\delta \geq 0$, be

$$p_k^\delta(\lambda) = 1 - a_1^\delta \lambda - \ldots - a_k^\delta \lambda^k\,;$$

note that these polynomials are well-defined and $a_k^\delta \neq 0$ for all sufficiently small $\delta \geq 0$. This yields

$$x_k^\delta := R_k y^\delta = \sum_{i=1}^k a_i^\delta T^{i-1} y^\delta\,.$$

The aim is to show that $a_i^\delta \to a_i^0$, $1 \leq i \leq n$, uniformly as $\delta \to 0$, which readily implies that R_k is continuous at $y^0 = y$.

Let $[\cdot, \cdot]_n^\delta$ be the inner product (2.5) defined by the right-hand side y^δ and let $\{\mu_m^\delta\}_{m \geq 0}$ be the corresponding moments, i.e.,

$$\mu_m^\delta = [1, \lambda^m]_n^\delta = \langle y^\delta, T^{m+n} y^\delta \rangle, \qquad m \in \mathbb{N}_0\,. \tag{2.28}$$

The orthogonality of $\{p_k^\delta\}$ with respect to $[\cdot, \cdot]_n^\delta$ yields

$$0 = [p_k^\delta, \lambda^m]_n^\delta = [1, \lambda^m]_n^\delta - \sum_{i=1}^k a_i^\delta [1, \lambda^{m+i}]_n^\delta = \mu_m^\delta - \sum_{i=1}^k \mu_{m+i}^\delta a_i^\delta, \qquad 0 \leq m \leq k-1\,.$$

Rewriting these k equations for $a_1^\delta, \ldots, a_k^\delta$ in matrix notation gives

$$M_k^\delta \mathbf{a}^\delta \equiv \begin{pmatrix} \mu_1^\delta & \mu_2^\delta & \cdots & \mu_k^\delta \\ \mu_2^\delta & \mu_3^\delta & & \mu_{k+1}^\delta \\ \vdots & & \ddots & \\ \mu_k^\delta & \mu_{k+1}^\delta & & \mu_{2k-1}^\delta \end{pmatrix} \begin{pmatrix} a_1^\delta \\ a_2^\delta \\ \vdots \\ a_k^\delta \end{pmatrix} = \begin{pmatrix} \mu_0^\delta \\ \mu_1^\delta \\ \vdots \\ \mu_{k-1}^\delta \end{pmatrix}.$$

It follows from (2.28) that $\mu_m^\delta \to \mu_m^0$, $0 \leq m \leq 2k-1$, as $\delta \to 0$, hence

$$M_k^\delta \longrightarrow M_k = \begin{pmatrix} \mu_1^0 & \mu_2^0 & \cdots & \mu_k^0 \\ \mu_2^0 & \mu_3^0 & & \mu_{k+1}^0 \\ \vdots & & \ddots & \\ \mu_k^0 & \mu_{k+1}^0 & & \mu_{2k-1}^0 \end{pmatrix}, \qquad \delta \to 0\,;$$

the convergence is uniform in $\|y - y^\delta\|$. To show that the moment matrix M_k is nonsingular, observe that for $\mathbf{z} = (\zeta_1, \cdots, \zeta_k)^T$,

$$z^* M_k z = \| T^{(n+1)/2} \sum_{i=1}^{k} \zeta_i T^{i-1} y \|^2 .$$

In other words, $z^* M_k z = 0$, if and only if

$$(\sum_{i=0}^{k-1} \zeta_{i+1} T^i) y \in \mathcal{N}(T) ,$$

that is, if and only if Py belongs to a $(k-1)$-dimensional invariant subspace of T. By assumption, this is not the case and therefore M_k is positive definite and has a continuous inverse M_k^{-1}. Consequently,

$$\mathbf{a}^\delta \longrightarrow \mathbf{a}^0 = M_k^{-1} \begin{pmatrix} \mu_0^0 \\ \vdots \\ \mu_{k-1}^0 \end{pmatrix} , \qquad \delta \to 0 ,$$

uniformly in $\| y - y^\delta \|$, as was to be shown.

The proof of the *if-part* is given for $y \in \mathcal{R}(T)$ first. Let κ be the dimension of the Krylov space $\mathcal{K}_k(y; T)$, and assume that $\kappa < k$. Then there exists a polynomial $p_\kappa \in \Pi_\kappa^0$ with

$$p_\kappa(T)y = 0 , \qquad p_\kappa(\lambda) = \prod_{j=1}^{\kappa} (1 - \lambda/\lambda_j) ,$$

where $\{\lambda_j\}_{j=1}^{\kappa}$ are mutually distinct positive numbers. Since (1.1) is ill-posed, the spectrum of T accumulates at $\lambda = 0$, and it is possible to choose $\lambda_{\kappa+1}, \ldots, \lambda_k$ from the spectrum of T with

$$0 < \lambda_k < \lambda_{k-1} < \cdots < \lambda_{\kappa+1} < \min_{j=1,\ldots,\kappa} \lambda_j .$$

For each $j \in \{\kappa+1, \ldots, k\}$ and each sufficiently small $\varepsilon > 0$ there exists $y_j^\varepsilon \in \mathcal{X}$ with $\| y_j^\varepsilon \| = 1$ and

$$E_{\lambda_j - \varepsilon} y_j^\varepsilon = 0 , \qquad E_{\lambda_j + \varepsilon} y_j^\varepsilon = y_j^\varepsilon .$$

Let

$$y^{\delta,\varepsilon} = y + \delta \sum_{j=\kappa+1}^{k} y_j^\varepsilon ,$$

and denote by $[\cdot, \cdot]_n^{\delta,\varepsilon}$ the inner product (2.5) induced by the right-hand side $y^{\delta,\varepsilon}$. If $\{\mu_m^{\delta,\varepsilon}\}_{m \geq 0}$ are the moments corresponding to $[\cdot, \cdot]_n^{\delta,\varepsilon}$ then it is easy to see that

$$\mu_m^{\delta,\varepsilon} = [1, \lambda^m]_n^{\delta,\varepsilon} = [1, \lambda^{m+n}]_0^{\delta,\varepsilon} \longrightarrow [1, \lambda^{m+n}]_0 + \delta^2 \sum_{j=\kappa+1}^{k} \lambda_j^{m+n} , \qquad \varepsilon \to 0 ,$$

for every $0 \leq m \leq 2k - 1$. This implies, as in the first part of this proof, that the corresponding residual polynomial $p_k^{\delta,\varepsilon}$ converges (uniformly) to the polynomial $p_k^\delta \in \Pi_k^0$ which is perpendicular to Π_{k-1} in the inner product space defined by

$$[\varphi, \psi]_n^\delta := [\varphi, \psi]_n + \delta^2 \sum_{j=\kappa+1}^{k} \lambda_j^n \varphi(\lambda_j) \psi(\lambda_j), \qquad \varphi, \psi \in \Pi_k.$$

Since the support of the distribution corresponding to this inner product is $\{\lambda_1, \ldots, \lambda_k\}$, p_k^δ has precisely these k points as roots. It follows that the kth iteration polynomial $q_{k-1}^{\delta,\varepsilon}$ corresponding to $p_k^{\delta,\varepsilon}$ via (2.2) satisfies

$$q_{k-1}^{\delta,\varepsilon}(\lambda_k) = \frac{1 - p_k^{\delta,\varepsilon}(\lambda_k)}{\lambda_k} \longrightarrow \frac{1}{\lambda_k}, \qquad \varepsilon \to 0,$$

and hence, for ε sufficiently small,

$$\|R_k y^{\delta,\varepsilon}\| = \|q_{k-1}^{\delta,\varepsilon}(T) y^{\delta,\varepsilon}\| \geq \frac{1}{2\lambda_k} \|E_{\lambda_k+\varepsilon} y^{\delta,\varepsilon}\| = \frac{1}{2\lambda_k} \delta. \tag{2.29}$$

As the spectrum of T clusters at $\lambda = 0$, λ_k can be made arbitrarily small. It follows that R_k is unbounded (and thus discontinuous) in any neighborhood of y.

This completes the proof for $y \in \mathcal{R}(T)$. If $y \notin \mathcal{R}(T)$ then a nontrivial component of y belongs to the orthogonal complement of $\mathcal{R}(T)$. By the discussion in Section 2.1 CG breaks down in the $\kappa + 1$st step; hence, R_k is undefined for CG. For conjugate gradient type methods with parameter $n \geq 1$, on the other hand,

$$P R_k y = R_k P y,$$

since the residual polynomials are not affected by any component of y in the orthogonal complement of $\mathcal{R}(T)$. Thus, R_k is discontinuous at y if it is unbounded near Py; since $Py \in \mathcal{R}(T)$ this has been established above and the proof is done. $\square$

Remark. The same result applies for the conjugate gradient type methods of Section 2.3, except that one has to consider invariant subspaces of TT^*. The proof is essentially the same. In the *if-part* the final estimate (2.29) becomes

$$\|R_k y^{\delta,\varepsilon}\| = \|q_{k-1}^{\delta,\varepsilon}(T^*T) T^* y^{\delta,\varepsilon}\| \geq \frac{1}{2\sqrt{\lambda_k}} \delta,$$

but this does not affect the conclusion.

Note that actually a stronger result has been proven, namely if R_k is continuous at $y \in \mathcal{X}$, then continuity is locally Lipschitz, i.e.,

$$\|R_k y - R_k y^\delta\| = O(\|y - y^\delta\|),$$

uniformly for y^δ sufficiently close to y.

Reconsider the situation encountered in the proof of Proposition 2.10, where

$$x = \sum_{j=1}^{k-1} \xi_j v_j, \qquad y = Tx,$$

with T compact, selfadjoint and positive definite, and $Tv_j = \lambda_j v_j$ for $j \in \mathbf{N}$. It has been shown that $x_k^\delta = R_k y^\delta$ diverges to infinity as $y^\delta \to y$ if T is non-degenerate and the perturbations $\{y^\delta\}$ are suitably chosen. On the other hand, if $\xi_j \neq 0$, $1 \leq j < k$, and if all eigenvalues λ_j, $1 \leq j < k$, are nonzero and mutually different, then it follows from Theorem 2.11 that

$$x_{k-1}^\delta = R_{k-1} y^\delta \longrightarrow R_{k-1} y = x, \qquad y^\delta \to y.$$

In other words, the $k-1$st iterate of the conjugate gradient type method is a regularized approximation of x in this particular instance. Any regularizing stopping rule *must* terminate the iteration with $k(y^\delta, \delta) = k - 1$ if y is as above and δ is sufficiently small.

Notes and remarks

Section 2.1. Forty years have passed since HESTENES and STIEFEL [45] developed CG and CGNE, forty years with very many contributions to the theory, cf., e.g., the bibliography by GOLUB and O'LEARY [25]. HAYES [43] extended CG to well-posed operator equations, and KAMMERER and NASHED [46] were probably the first to analyze CGNE for ill-posed problems. Conjugate gradient methods with respect to general inner products including the present ones as special cases have been considered by several authors; the term "conjugate gradient type method" is the same used, e.g., by LARDY [51].

Section 2.2. The MR-method goes back to a paper by LANCZOS [50] on eigenvalue approximations (see also STIEFEL [74]), while the name is adopted from the implementation MINRES by PAIGE and SAUNDERS [64]. The name "conjugate residual method" can also be found in the literature. GILYAZOV [22] was the first to study MR for selfadjoint, ill-posed problems.

Section 2.3. While CGNE has already been introduced by HESTENES and STIEFEL [45], the minimal error method CGME occurs first in papers by CRAIG [11] and SHAMANSKII [73]. KING [47] called it "minimal error method" and obtained convergence rates for ill-posed problems with exact data $y \in \mathcal{R}(T)$, see also LARDY [51].

Section 2.4. Most of the results of this section are well known, although the proofs as given here are somewhat nonstandard. In using this particular perspective, the intimate relation between conjugate gradient type methods and orthogonal polynomials is especially emphasized. STIEFEL [74] was the first to elaborate on this point of view; see also FISCHER [18]. Propositions 2.5 and 2.8 are intrinsic in several works; GUTKNECHT [31] provides further relations between "consecutive" conjugate gradient type methods.

Section 2.5. Algorithm 2.6 is taken from a paper by LARDY [51]. GUTKNECHT [31] gives equivalent formulas for computing the iterates of "adjacent" conjugate gradient type methods.

Section 2.6. The effect of round-off on the conjugate gradient iteration was studied analytically by PAIGE [63], GREENBAUM [27], and DRUSKIN and KNIZHNERMAN [12]; a number of illuminating numerical experiments can be found in [75, 28, 60].

The (dis)continuity of conjugate gradient type iterations was investigated by EICKE, LOUIS and PLATO [13] for compact operator equations; Proposition 2.10 is due to them.

3. Regularizing Properties of MR and CGNE

This chapter presents the analysis of conjugate gradient type methods with parameter $n \geq 1$ including, e.g., MR and CGNE. It will be shown that the iterates converge monotonically to $T^\dagger y$ if the right-hand side data y belong to $\mathcal{R}(T)$. If $y \notin \mathcal{R}(T)$ then the iterates typically diverge to infinity in norm. Nevertheless, if the right-hand side y^δ is an approximation of $y \in \mathcal{R}(T)$, then some iterates approximate the exact solution $T^\dagger y$ with order-optimal accuracy. The crux of the matter is to decide when this is the case. If the magnitude of the perturbation $\|y - y^\delta\|$ is known, such a decision can be based on the discrepancy principle. Otherwise, heuristic arguments are required to halt the iteration. One heuristic stopping rule is presented at the end of this chapter; it has the nice feature that it provides an a posteriori error estimate for the corresponding approximation.

3.1 Monotonicity, convergence and divergence

The aim of this first section is to investigate the qualitative behavior of the iterates as $k \to \infty$. It will be shown that for data $y \in \mathcal{R}(T)$ the iterates $\{x_k\}$ converge to $T^\dagger y$, and the convergence is monotone. The key ingredient to the proof of this result is a simple consequence of Lemma 2.4:

Lemma 3.1 *Let $y \in \mathcal{R}(T)$, and let m and n be integers with $m < n$ and such that $[1,1]_m < \infty$. Then, given any fixed k with $0 \leq k \leq \kappa$, all expansion coefficients γ_j, $0 \leq j \leq k$, in*

$$p_k^{[n]} = \gamma_0 p_0^{[m]} + \gamma_1 p_1^{[m]} + \ldots + \gamma_k p_k^{[m]}$$

are nonnegative. If $k \neq \kappa$ then all coefficients are strictly positive.

Proof. Note that m and n may be negative but the assumption that $[1,1]_m < \infty$ guarantees that all orthogonal polynomials are well defined. Recall that for $k = \kappa$ the polynomials $p_\kappa^{[n]}$ coincide for all admissible values of n since $y \in \mathcal{R}(T)$, and hence, all expansion coefficients except for $\gamma_\kappa = 1$ vanish in this case. For $k < \kappa$ the proof goes by induction on $n - m$. Assume first that $n - m = 1$, i.e., $m = n - 1$: if $n > 0$ then the nonnegativity of the expansion coefficients follows immediately from their explicit representation given in Lemma 2.4. As discussed in the remark following Lemma 2.4 this derivation remains valid for $n \leq 0$ since $y \in \mathcal{R}(T)$ holds by assumption. Having now established the assertion for $m = n - 1$ one can insert the respective nonnegative expansions for $p_j^{[n-1]}$, $0 \leq j \leq k$, into the right-hand side of (2.20) to obtain a non-negative expansion for $m = n - 2$, and so on. $\qquad\square$

This lemma can be used to prove the following theorem.

Theorem 3.2 *Let $y \in \mathcal{R}(T)$, and let m and n be integers with $m \leq n - 1$ and $[1,1]_m < \infty$. Then $[p_k^{[n]}, p_k^{[n]}]_m$ is strictly decreasing as k goes from 0 to κ.*

Proof. Fix $0 \leq k < \kappa$, and consider

$$\Delta := [p_k^{[n]}, p_k^{[n]}]_m - [p_{k+1}^{[n]}, p_{k+1}^{[n]}]_m \, .$$

The aim is to show that Δ is positive. From Proposition 2.5 follows

$$\Delta = [p_k^{[n]} - p_{k+1}^{[n]}, p_k^{[n]} + p_{k+1}^{[n]}]_m = \pi_{k,n}[p_k^{[n+1]}, p_k^{[n]} + p_{k+1}^{[n]}]_{m+1} \qquad (3.1)$$

with $\pi_{k,n} > 0$. In the case when $m = n - 1$ (3.1) becomes by orthogonality

$$
\begin{aligned}
\Delta &= \pi_{k,n}[p_k^{[n+1]}, p_k^{[n]}]_n \\
&= \pi_{k,n}[p_k^{[n+1]}, p_k^{[n+1]}]_n + \pi_{k,n}[p_k^{[n+1]}, \frac{p_k^{[n]} - p_k^{[n+1]}}{\lambda}]_{n+1} \\
&= \pi_{k,n}[p_k^{[n+1]}, p_k^{[n+1]}]_n \, ,
\end{aligned}
$$

which is positive. In the case when $m < n - 1$ Lemma 3.1 will be applied as follows. Expanding $p_k^{[n+1]}$ and $p_k^{[n]} + p_{k+1}^{[n]}$ on the right-hand side of (3.1) in terms of $\{p_k^{[m+1]}\}$ yields positive expansion coefficients γ_i $(0 \leq i \leq k)$ and $\tilde{\gamma}_j$ $(0 \leq j \leq k+1)$, respectively, and hence,

$$\Delta = \pi_{k,n} \sum_{i=0}^{k} \sum_{j=0}^{k+1} \gamma_i \tilde{\gamma}_j [p_i^{[m+1]}, p_j^{[m+1]}]_{m+1} = \pi_{k,n} \sum_{j=0}^{k} \gamma_j \tilde{\gamma}_j [p_j^{[m+1]}, p_j^{[m+1]}]_{m+1} \, .$$

Again, the right-hand side is positive proving thus the monotonicity of $[p_k^{[n]}, p_k^{[n]}]_m$ with respect to k. $\qquad\square$

Remark. In general, Theorem 3.2 does not extend to $m \geq n$; compare Example 4.3 in Section 4.1 for a counterexample showing that $[p_k^{[n]}, p_k^{[n]}]_n$ does not decrease monotonically.

Two important consequences of this theorem shall be mentioned explicitly. One of these is the (not very well known) monotonicity of the iteration error of MR and CGNE.

36

Corollary 3.3 *Let $\{x_k\}$ denote the iterates of a conjugate gradient type method with parameter $n \geq 1$ and right-hand side $y \in \mathcal{R}(T)$. Then the following holds.*

(i) *The residual norm $\|y - Tx_k\|$ is strictly decreasing for $0 \leq k \leq \kappa$.*

(ii) *The iteration error $\|T^\dagger y - x_k\|$ is strictly decreasing for $0 \leq k \leq \kappa$.*

Proof. Recall that $\|y - Tx_k\|^2 = [p_k, p_k]_0$ and $\|T^\dagger y - x_k\|^2 = [p_k, p_k]_{-2}$. Since $p_k = p_k^{[n]}$ with $n \geq 1$ the assertions follow from Theorem 3.2. $\qquad\square$

With this corollary available it is comparatively easy to establish the following convergence result.

Theorem 3.4 *If the right-hand side y belongs to $\mathcal{R}(T)$ then the iterates $\{x_k\}$ of a conjugate gradient type method with parameter $n \geq 1$ converge to $T^\dagger y$ as $k \to \infty$.*

Proof. If $\kappa < \infty$ then $p_\kappa(T)y = 0$ according to the discussion in Section 2.1, and hence the conjugate gradient type method terminates after κ steps with $x_\kappa = T^\dagger y$. Assume next that $\kappa = \infty$. From Proposition 2.1 follows that

$$\|T^{(n-1)/2}(y - Tx_k)\|^2 \leq [\varphi_k, \varphi_k]_{n-1}$$

for any $\varphi_k \in \Pi_k^0$. Taking $\varphi_k(\lambda) = (1-\lambda)^k$, the right-hand side $[\varphi_k, \varphi_k]_{n-1}$ converges to $\|E_0 T^{(n-1)/2}y\|^2$ by the Banach-Steinhaus theorem, since $\{\varphi_k\}$ is uniformly bounded and converges pointwise to the zero function on $(0, 1]$. As $E_0 y = (I - P)y = 0$ it follows that

$$T^{(n-1)/2}(y - Tx_k) \to 0, \qquad k \to \infty. \tag{3.2}$$

Let x be the solution of $Tx = y$ in $\mathcal{R}(T)$ so that

$$\|x - x_k\|^2 = \|p_k(T)x\|^2 = [p_k, p_k]_{-2} \cdot$$

As stated in Corollary 3.3, $\|x - x_k\|$ is monotonically decreasing, and hence the sequence $\{x - x_k\}$ is bounded. Let z be a weak limit of some subsequence of $\{x - x_k\}$. Clearly, $z \perp \mathcal{N}(T)$, and

$$T^{(n-1)/2}(y - Tx_k) = T^{(n+1)/2}(x - x_k) \rightharpoonup T^{(n+1)/2}z$$

for the corresponding indices k going to infinity. In view of (3.2) this implies that $z = 0$, proving that the entire sequence $\{x - x_k\}$ converges weakly to zero as $k \to \infty$. From this follows that

$$[p_k, p_0]_{-2} = \langle p_k(T)x, x \rangle = \langle x - x_k, x \rangle \longrightarrow 0, \qquad k \to \infty. \tag{3.3}$$

From Proposition 2.5 one has

$$[p_k, p_k]_{-2} = [p_k, p_0]_{-2} + \sum_{j=0}^{k-1} [p_k, p_{j+1} - p_j]_{-2} = [p_k, p_0]_{-2} - \sum_{j=0}^{k-1} \pi_{j,n} [p_k, p_j^{[n+1]}]_{-1} ,$$

and Lemma 3.1 therefore yields

$$[p_k, p_k]_{-2} = [p_k, p_0]_{-2} - \sum_{j=0}^{k-1} \pi_{j,n} \sum_{i=1}^{j} \gamma_i \tilde{\gamma}_{i,j} [p_i^{[-1]}, p_i^{[-1]}]_{-1}$$

with positive expansion coefficients γ_i and $\tilde{\gamma}_{i,j}$. Since $\pi_{j,n}$ is positive this establishes that

$$[p_k, p_k]_{-2} < [p_k, p_0]_{-2} .$$

From (3.3) now follows that $[p_k, p_k]_{-2} = \|x - x_k\|^2$ goes to zero as $k \to \infty$ as was to be shown. □

Remark. Concerning the algorithms of Section 2.3 with parameters $n \geq 1$ (including CGNE) Theorem 3.4 remains true for $y \in \mathcal{D}(T^\dagger)$. To see this one first observes that the iterates x_k do not depend on the component of $y \in \mathcal{D}(T^\dagger) \setminus \mathcal{R}(T)$, and one can therefore assume without loss of generality that $y \in \mathcal{R}(T)$. Then the proof as given above extends readily to this setting since for $y \in \mathcal{R}(T)$ one has

$$\|T^\dagger y - x_k\|^2 = \|p_k(T^*T)T^\dagger y\|^2 = [p_k, p_k]_{-1} .$$

Even when $y \in \mathcal{R}(T)$, arbitrarily small perturbations of y - caused by measurement errors, say - need no longer belong to $\mathcal{R}(T)$ as the range of T is non-closed. In this case the iteration will diverge as the following result shows.

Theorem 3.5 *Consider a conjugate gradient type method with parameter $n \geq 1$ and right-hand side $y \notin \mathcal{R}(T)$. If $\kappa = \infty$ then $\|x_k\| \to \infty$ as $k \to \infty$. If $\kappa < \infty$ then the iteration terminates after κ steps; in this case $Tx_\kappa = Py$, and $x_\kappa = T^\dagger y$ if and only if $\kappa = 0$.*

Proof. Assume first that $Py \neq y$, i.e., $E_0 y \neq 0$. If $\kappa < \infty$ then the iteration terminates with $p_\kappa(T)y = y - Tx_\kappa = E_0 y$, and one has $(I - E_0)x_\kappa = T^\dagger y$. Consequently, $x_\kappa = T^\dagger y$ if and only if $E_0 x_\kappa = q_{\kappa-1}(0)E_0 y = 0$, i.e., if and only if $\kappa = 0$. If $\kappa = \infty$ then the iteration does not terminate and, in view of (2.19), $|p_k'(0)| \geq k$ for every $k \in \mathbb{N}$. Since $E_0 y \neq 0$ this implies

$$\|x_k\| \geq q_{k-1}(0)\|E_0 y\| = |p_k'(0)| \, \|E_0 y\| \longrightarrow \infty, \qquad k \to \infty .$$

It remains to consider the case when $Py = y$. Since $y \notin \mathcal{D}(T^\dagger)$ one has $\kappa = \infty$ in this case, and the iteration does not terminate. As in the proof of Theorem 3.4

38

let $\varphi_k(\lambda) = (1 - \lambda)^k \in \Pi_k^0$. By the Banach-Steinhaus theorem $\varphi_k(T)y \to 0$, and Proposition 2.1 implies that

$$\limsup_{k \to \infty} \|T^{(n-1)/2}(y - Tx_k)\| \leq \lim_{k \to \infty} \|T^{(n-1)/2}\varphi_k(T)y\| = 0. \qquad (3.4)$$

Assume now that some subsequence of $\{x_k\}$ remains bounded, so that it has a weakly converging subsequence with weak limit x, say. It follows that the images of this subsequence converge weakly to Tx and, because of (3.4), this yields $Tx = y$, i.e., $y \in \mathcal{R}(T)$ which contradicts the assumptions of the theorem. $\qquad \square$

Remark. If the conjugate gradient type method is applied with TT^* instead of T, then x_k converges to $T^\dagger y$ whenever $y \in \mathcal{D}(T^\dagger)$ as has been shown in the remark following Theorem 3.4. In the remaining case, $y \notin \mathcal{D}(T^\dagger)$, the iterates diverge to infinity in norm. The proof is the same as the one for the second case $Py = y$ as given above.

As a consequence, conjugate gradient type methods will in general lead to numerical instabilities if too many steps are performed with perturbed data $y^\delta \notin \mathcal{R}(T)$. Instead, the iteration has to be terminated appropriately: if the unperturbed right-hand side belongs to a finite dimensional invariant subspace of T, then this has already been exemplified in a remark following Theorem 2.11; in the remaining case, stability of a fixed number of iterations (Theorem 2.11) and convergence for unperturbed data (Theorem 3.4) in principle imply the existence of a regularizing stopping rule, cf. [78]. One such stopping rule is considered in Section 3.3.

3.2 Convergence rate estimates

There are many bounds for the rate of convergence when equation (1.1) is well-posed, i.e., when the spectrum of the selfadjoint, positive definite operator T is contained in a proper subinterval $[a, b]$ of $\mathbf{R}^+$, cf., e.g., [32]: in this case the iterates $\{x_k\}$ converge to the unique solution x of (1.1) with at least linear rate of convergence, and a standard upper bound for the logarithmic convergence factor can be obtained from the condition number of T, i.e., $\operatorname{cond} T = \|T\| \|T^{-1}\|$. The convergence factor approaches 1 as $\operatorname{cond} T \to \infty$, and the convergence rate slows down.

For ill-posed problems one can always construct data $y \in \mathcal{R}(T)$ such that the rate of convergence $x_k \to T^\dagger y$ is arbitrarily slow, cf. Lemma 5.3 (iii) with $n = -2$ and ν arbitrarily close to zero. With additional assumptions, however, sublinear convergence rates can be established. The main assumption that is used throughout the mathematical literature is the following:

Assumption 3.6 *The solution $x = T^\dagger y$ of the unperturbed right-hand side $y \in \mathcal{R}(T)$ belongs to $\mathcal{R}(|T|^\mu)$ for some $\mu > 0$, i.e., there exists $w \in \mathcal{X}$ with $x = (T^*T)^{\mu/2} w$; for brevity, let $\omega = \|w\|$.*[*]

As $T^\dagger$ is unbounded the range of T is non-closed, hence the sets $\mathcal{R}(|T|^\mu)$ form a decreasing scale of Hilbert spaces, where the topologies are defined via the corresponding preimages, respectively. Accordingly, the induced norms are related through the so-called *interpolation inequality* (cf., e.g., [81]):

$$\|T^\sigma x\| \leq \|x\|^{1-\sigma/\tau} \|T^\tau x\|^{\sigma/\tau}, \qquad 0 < \sigma < \tau. \tag{3.5}$$

It is clear that Assumption 3.6 represents some kind of a priori information. In many instances, e.g., for Fredholm integral equations of the first kind, the operator T is smoothing, and therefore Assumption 3.6 postulates a certain degree of smoothness of the exact solution. For example, if (1.1) is the problem of differentiation, i.e., $x = y'$, then $x \in \mathcal{R}(|T|^\mu)$ implies that the solution itself belongs to a Sobolev space of order μ. A posteriori bounds for the rate of convergence of conjugate gradient type methods with exact data y satisfying Assumption 3.6 can be found in Corollary 3.9.

The analysis of conjugate gradient type methods is substantially more complicated than, for example, the analysis of Tikhonov regularization. The reason is that on the one hand, the kth residual polynomial of a conjugate gradient type method depends on the data y^δ, i.e., R_k is a *nonlinear* operator; in Tikhonov regularization, for fixed regularization parameter ε, the function which takes the role of the kth residual polynomial is the rational $\varepsilon/(\lambda + \varepsilon)$, independent of the actual data y^δ, and the operator R_ε as given in (1.2) is linear. On the other hand, $\varepsilon/(\lambda + \varepsilon)$ is always bounded by 1 whatever $\varepsilon > 0$ is considered, whereas the residual polynomials of a conjugate gradient method need not be uniformly bounded over $[0,1]$. This hinders, for example, the application of the interpolation inequality (3.5) to obtain upper bounds for the iteration error from the norm of the residual with perturbed data. Instead, other – more technical – tools have to be employed.

In the following, denote by $\{x_k^\delta\}$ the iterates corresponding to y^δ, i.e., $x_k^\delta = R_k y^\delta$. The two Lemmas 3.7 and 3.8 below provide the central inequalities for the norms of the residual $y^\delta - T x_k^\delta$ and the error $T^\dagger y - x_k^\delta$, respectively. As can be seen from these bounds, the numbers $|p_k'(0)|$, $k \in \mathbb{N}$, essentially determine the rate of convergence and divergence of the iteration. Estimation of this "modulus of convergence" therefore amounts to an important task in the analysis of conjugate gradient type methods.

It must be emphasized that the inequalities provided by the two lemmas are not valid for conjugate gradient type methods which work with TT^* instead of T. Nevertheless, the proofs can readily be extended to cope with this different situation; the corresponding results are stated in subsequent remarks.

---

[*]In this form, Assumption 3.6 applies for selfadjoint *and* non-selfadjoint operators T.

40

Lemma 3.7 *Consider the iterates x_k^δ of a conjugate gradient type method with parameter $n \geq 1$ and right-hand side y^δ. If the exact right-hand side y belongs to $\mathcal{R}(T)$ and if $\kappa = \infty$ then*

$$\limsup_{k \to \infty} \|y^\delta - Tx_k^\delta\| \leq \|y - y^\delta\| .$$

If y satisfies Assumption 3.6 then one has

$$\|y^\delta - Tx_k^\delta\| \leq \|y - y^\delta\| + c \, |p_k'(0)|^{-\mu - 1} \omega, \qquad 1 \leq k \leq \kappa . \tag{3.6}$$

Proof. Let $\{\lambda_{j,k}\}_{j=0}^k$ denote the zeros of p_k again. Consequently, $p_k(\lambda)/(\lambda - \lambda_{1,k})$ is a polynomial of degree $k - 1$, and the orthogonality of $\{p_k\}$ gives

$$0 = [p_k, \frac{p_k}{\lambda - \lambda_{1,k}}]_n = \int_0^\infty p_k(\lambda) \, \frac{p_k(\lambda)}{\lambda - \lambda_{1,k}} \, \lambda^n \, d\|E_\lambda y^\delta\|^2 ,$$

or equivalently,

$$\int_0^{\lambda_{1,k}} p_k^2(\lambda) \, \frac{\lambda^n}{\lambda_{1,k} - \lambda} \, d\|E_\lambda y^\delta\|^2 = \int_{\lambda_{1,k}}^\infty p_k^2(\lambda) \, \frac{\lambda}{\lambda - \lambda_{1,k}} \lambda^{n-1} \, d\|E_\lambda y^\delta\|^2 .$$

Since $\lambda/(\lambda - \lambda_{1,k}) \geq 1$ for $\lambda \geq \lambda_{1,k}$ this yields

$$\int_0^{\lambda_{1,k}} p_k^2(\lambda) \, \frac{\lambda^n}{\lambda_{1,k} - \lambda} \, d\|E_\lambda y^\delta\|^2 \geq \int_{\lambda_{1,k}}^\infty p_k^2(\lambda) \lambda^{n-1} \, d\|E_\lambda y^\delta\|^2 .$$

Consequently,

$$\|y^\delta - Tx_k^\delta\|^2 = \int_0^{\lambda_{1,k}} p_k^2(\lambda) \, d\|E_\lambda y^\delta\|^2 + \int_{\lambda_{1,k}}^\infty p_k^2(\lambda) \, d\|E_\lambda y^\delta\|^2$$

$$\leq \int_0^{\lambda_{1,k}} p_k^2(\lambda) \, d\|E_\lambda y^\delta\|^2 + \lambda_{1,k}^{-(n-1)} \int_{\lambda_{1,k}}^\infty p_k^2(\lambda) \lambda^{n-1} \, d\|E_\lambda y^\delta\|^2$$

$$\leq \int_0^{\lambda_{1,k}} p_k^2(\lambda) \left(1 + (\frac{\lambda}{\lambda_{1,k}})^{n-1} \frac{\lambda}{\lambda_{1,k} - \lambda} \right) d\|E_\lambda y^\delta\|^2$$

$$\leq \int_0^{\lambda_{1,k}} p_k^2(\lambda) (1 + \frac{\lambda}{\lambda_{1,k} - \lambda}) \, d\|E_\lambda y^\delta\|^2 .$$

Introducing

$$\varphi_k(\lambda) := p_k(\lambda) \left(\frac{\lambda_{1,k}}{\lambda_{1,k} - \lambda} \right)^{1/2}, \qquad 0 \leq \lambda \leq \lambda_{1,k} , \tag{3.7}$$

the above inequality can be rewritten as

$$\|y^\delta - Tx_k^\delta\| \leq \|E_{\lambda_{1,k}} \varphi_k(T) y^\delta\| . \tag{3.8}$$

Note that φ_k^2 is bounded by 1 in $[0, \lambda_{1,k}]$. It therefore follows that

$$
\begin{aligned}
\|y^\delta - Tx_k^\delta\| &\leq \|E_{\lambda_{1,k}}\varphi_k(T)(y^\delta - y)\| + \|E_{\lambda_{1,k}}\varphi_k(T)y\| \\
&\leq \|y - y^\delta\| + \|E_{\lambda_{1,k}}\varphi_k(T)Tx\|.
\end{aligned}
\tag{3.9}
$$

Using elementary calculus it is easily seen from (3.7) and (2.19) that for $\nu > 0$ the maximum of $\lambda^\nu \varphi_k^2(\lambda)$ in $[0, \lambda_{1,k}]$ is attained at $\lambda = \lambda_*$, which is given as the unique solution in $(0, \lambda_{1,k})$ of

$$
\nu + \lambda_* \left(\frac{1}{\lambda_{1,k} - \lambda_*} - \sum_{j=1}^{k} \frac{2}{\lambda_{j,k} - \lambda_*} \right) = 0.
$$

It follows that

$$
\nu \geq \lambda_* \sum_{j=1}^{k} \frac{1}{\lambda_{j,k} - \lambda_*} \geq \lambda_* \sum_{j=1}^{k} \frac{1}{\lambda_{j,k}} = \lambda_* |p_k'(0)|,
$$

cf. (2.19). Thus, $\lambda_* \leq \nu |p_k'(0)|^{-1}$ and one obtains

$$
\lambda^\nu \varphi_k^2(\lambda) \leq \lambda_*^\nu \varphi_k^2(\lambda_*) \leq \nu^\nu |p_k'(0)|^{-\nu}, \qquad \nu > 0,\ 0 \leq \lambda \leq \lambda_{1,k}.
\tag{3.10}
$$

Inserting (3.10) with $\nu = 2$, (3.9) can be estimated as follows:

$$
\|y^\delta - Tx_k^\delta\| \leq \|y - y^\delta\| + 2|p_k'(0)|^{-1} \|x\|.
$$

Thus, since $|p_k'(0)| \to \infty$ as $k \to \infty$, the assertion on the limit superior of $\|y^\delta - Tx_k^\delta\|$ follows.

If Assumption 3.6 is satisfied, i.e., if $x = T^\mu w$ with $\|w\| = \omega$, then one proceeds in the same way: (3.9) and (3.10) with $\nu = 2\mu + 2$ yield

$$
\|y^\delta - Tx_k^\delta\| \leq \|y - y^\delta\| + \|E_{\lambda_{1,k}}\varphi_k(T)T^{\mu+1}w\| \leq \|y - y^\delta\| + (2\mu+2)^{\mu+1}|p_k'(0)|^{-\mu-1}\omega,
$$

and the proof is complete. $\qquad\qquad\qquad\qquad\qquad\qquad\qquad\qquad\qquad\square$

Remark. Under the assumptions of Lemma 3.7 the following estimate for conjugate gradient type methods working with TT^* instead of T can be proven in a similar way:

$$
\|y^\delta - Tx_k^\delta\| \leq \|y - y^\delta\| + c\,|p_k'(0)|^{-(\mu+1)/2}\omega.
$$

To prove this one may assume without loss of generality that y^δ lies in the closure of $\mathcal{R}(T)$. Then the proof is very similar to the one given above: replacing T by TT^* everywhere, one obtains

$$
\|y^\delta - Tx_k^\delta\| \leq \|E_{\lambda_{1,k}}\varphi_k(TT^*)y^\delta\|;
$$

To obtain the desired inequalities one only has to use different values of ν in (3.10), e.g., $\nu = \mu + 1$ instead of $2\mu + 2$.

The next result estimates the error $T^\dagger y - x_k^\delta$.

Lemma 3.8 *Consider a conjugate gradient type method with parameter $n \geq 1$ and right-hand side y^δ. Assume that y satisfies Assumption 3.6. Then, for $0 \leq k \leq \kappa$,*

$$\|T^\dagger y - x_k^\delta\| \leq c\left(\omega^{1/\mu+1}\rho_k^{\mu/\mu+1} + |p_k'(0)|\,\|y - y^\delta\|\right), \qquad (3.11)$$

where

$$\rho_k := \max\{\|y^\delta - Tx_k^\delta\|, \|y - y^\delta\|\}. \qquad (3.12)$$

Proof. Consider first the case $k = 0$. By virtue of Assumption 3.6 the interpolation inequality (3.5) yields

$$\|T^\dagger y\| \leq \|w\|^{1/\mu+1}\|y\|^{\mu/\mu+1}.$$

Since $\|y\| \leq \|y^\delta\| + \|y - y^\delta\| \leq 2\rho_0$ by the definition (3.12) of ρ_0, this yields

$$\|T^\dagger y\| \leq 2^{\mu/\mu+1}\,\omega^{1/\mu+1}\rho_0^{\mu/\mu+1},$$

proving the validity of (3.11) for $k = 0$.

For $k > 0$ let ε be such that

$$0 < \varepsilon \leq |p_k'(0)|^{-1}, \qquad (3.13)$$

which in particular implies that ε is smaller than $\lambda_{1,k}$, cf. (2.19). Next, introduce $x = T^\dagger y$ and

$$\tilde{x}_k = q_{k-1}(T)y;$$

it follows that $x - \tilde{x}_k = p_k(T)x$. Note that $\tilde{x}_k$ is *not* the iterate that were computed by the conjugate gradient type algorithm with exact right-hand side y, since in general this would result in a different polynomial q_{k-1}. Using $\tilde{x}_k$ one obtains

$$
\begin{aligned}
\|x - x_k^\delta\| &\leq \|E_\varepsilon(x - x_k^\delta)\| + \|(I - E_\varepsilon)(x - x_k^\delta)\| \\
&\leq \|E_\varepsilon(x - \tilde{x}_k)\| + \|E_\varepsilon(\tilde{x}_k - x_k^\delta)\| + \varepsilon^{-1}\|(I - E_\varepsilon)(y - Tx_k^\delta)\| \\
&\leq \|E_\varepsilon p_k(T)T^\mu w\| + \|E_\varepsilon q_{k-1}(T)(y - y^\delta)\| + \varepsilon^{-1}\|y - Tx_k^\delta\| \\
&\leq \|\lambda^\mu p_k(\lambda)\|_{[0,\varepsilon]}\,\omega + \|q_{k-1}(\lambda)\|_{[0,\varepsilon]}\,\|y - y^\delta\| \\
&\qquad\qquad + \varepsilon^{-1}(\|y^\delta - Tx_k^\delta\| + \|y - y^\delta\|).
\end{aligned}
$$

Since $\varepsilon < \lambda_{1,k}$, p_k is convex in $[0,\varepsilon]$ by (2.19), and hence

$$0 \leq q_{k-1}(\lambda) = \frac{1 - p_k(\lambda)}{\lambda} \leq |p_k'(0)| \,, \qquad 0 \leq \lambda \leq \varepsilon. \tag{3.14}$$

Furthermore,

$$0 \leq \lambda^\mu p_k(\lambda) \leq \varepsilon^\mu \,, \qquad 0 \leq \lambda \leq \varepsilon.$$

This, together with the definition (3.12) of ρ_k, results in the following bound:

$$\|x - x_k^\delta\| \leq \varepsilon^\mu \omega + 2\varepsilon^{-1}\rho_k + |p_k'(0)| \, \|y - y^\delta\|. \tag{3.15}$$

Note that the right-hand side of (3.15) is a decreasing function of ε in $(0, \varepsilon_*)$ and increasing in (ε_*, ∞), where ε_* is determined through

$$\varepsilon_*^{\mu+1} = \frac{2}{\mu} \frac{\rho_k}{\omega}.$$

Taking (3.13) into account the estimation of (3.15) splits into two different cases as follows: first, if $\varepsilon_* \geq |p_k'(0)|^{-1}$ then let $\varepsilon = |p_k'(0)|^{-1}$ in (3.15), which yields

$$\|x - x_k^\delta\| \leq |p_k'(0)|^{-\mu}\omega + 2|p_k'(0)|\rho_k + |p_k'(0)| \, \|y - y^\delta\|.$$

Lemma 3.7 and the definition (3.12) of ρ_k imply that

$$\rho_k \leq \|y - y^\delta\| + c|p_k'(0)|^{-\mu-1}\omega.$$

Combining these two estimates for the case when $\varepsilon_* \geq |p_k'(0)|^{-1}$ it follows that

$$\begin{aligned}
\|x - x_k^\delta\| &\leq c|p_k'(0)|^{-\mu}\omega + 3|p_k'(0)| \, \|y - y^\delta\| \\
&\leq c\varepsilon_*^\mu\omega + 3|p_k'(0)| \, \|y - y^\delta\| \\
&= c\omega^{1/\mu+1}\rho_k^{\mu/\mu+1} + 3|p_k'(0)| \, \|y - y^\delta\|.
\end{aligned}$$

On the other hand, if $\varepsilon_* < |p_k'(0)|^{-1}$ then one can choose $\varepsilon = \varepsilon_*$ in (3.15) to obtain

$$\begin{aligned}
\|x - x_k^\delta\| &\leq \varepsilon_*^\mu\omega + 2\varepsilon_*^{-1}\rho_k + |p_k'(0)| \, \|y - y^\delta\| \\
&= c\omega^{1/\mu+1}\rho_k^{\mu/\mu+1} + |p_k'(0)| \, \|y - y^\delta\|.
\end{aligned}$$

Thus, in any case (3.11) holds true. □

Combining the two lemmas one obtains the following result which contains the aforementioned bound for the convergence rate of conjugate gradient type methods for ill-posed problems with exactly given right-hand side $y \in \mathcal{R}(T)$ as a special case.

44

Corollary 3.9 *If y satisfies Assumption 3.6 and $\|y - y^\delta\| \leq \delta$, then the iteration error of conjugate gradient type methods with parameter $n \geq 1$ is bounded by*

$$\|T^\dagger y - x_k^\delta\| \leq c\left(|p_k'(0)|^{-\mu}\omega + |p_k'(0)|\delta\right), \qquad 1 \leq k \leq \kappa. \tag{3.16}$$

Proof. Inserting (3.6) into (3.11) one obtains

$$\|T^\dagger y - x_k^\delta\| \leq c\left(|p_k'(0)|^{-\mu}\omega + |p_k'(0)|\delta + \omega^{1/\mu+1}\delta^{\mu/\mu+1}\right), \qquad k \geq 1.$$

Since there is a constant $c > 0$ such that

$$\omega^{1/\mu+1}\delta^{\mu/\mu+1} \leq c(t^{-\mu}\omega + t\delta) \qquad \text{for every } t > 0,$$

(3.16) follows. $\qquad\qquad\qquad\qquad\qquad\qquad\qquad\qquad\qquad\qquad\qquad\qquad\qquad\qquad\qquad\quad\square$

A discussion of this result is necessary. While the first term on the right-hand side of (3.16) goes to zero as $k \to \infty$, the second term will diverge to infinity. For smaller values of k, however, this diverging term is of the order of δ only, and therefore the iterates seem to converge in the beginning: simply speaking, the iteration process does not "see" any data perturbation. After a while, both terms on the right-hand side of (3.16) have about the same order of magnitude, and from that point onwards the propagated data error dominates the total error. The iteration diverges. This effect is called *semiconvergence*, and can nicely be seen in numerical examples, e.g., in Figures 5.4 and 6.7.

Remark. The error of conjugate gradient type methods working with TT^* instead of T can be estimated with much the same technique. There is only one additional difficulty, as it becomes necessary to estimate $\lambda^{1/2}q_{k-1}(\lambda)$ instead of $q_{k-1}(\lambda)$, cf. (3.14). The required estimation goes as follows:

$$0 \leq \lambda q_{k-1}^2(\lambda) = \frac{1 - p_k(\lambda)}{\lambda}(1 - p_k(\lambda)) \leq |p_k'(0)|, \qquad 0 \leq \lambda \leq \varepsilon.$$

Eventually, this leads to the following bound for the iteration error, valid under the same assumptions as in Lemma 3.8:

$$\|T^\dagger y - x_k^\delta\| \leq c\left(\omega^{1/\mu+1}\rho_k^{\mu/\mu+1} + |p_k'(0)|^{1/2}\|y - y^\delta\|\right),$$

Corollary 3.9 holds accordingly with $|p_k'(0)|$ replaced everywhere by $|p_k'(0)|^{1/2}$.

3.3 The discrepancy principle

In 1966, MOROZOV [56] suggested the *discrepancy principle* as a means for choosing the regularization parameter in Tikhonov regularization. Roughly speaking, his idea

was to put up with an error of magnitude δ in the data fit of the computed approximation, to prevent unwanted magnification of noise components of the right-hand side.
In the context of iteration methods, the discrepancy principle defines the following
stopping rule.

Stopping Rule 3.10 (Discrepancy principle) *Assume* $\|y - y^\delta\| \le \delta$. *Fix* $\tau > 1$,
and terminate the iteration when, for the first time, $\|y^\delta - Tx_k^\delta\| \le \tau\delta$. *Denote by* $k(\delta)$
the resulting stopping index[*].

According to Lemma 3.7 the discrepancy principle always determines a finite stop-
ping index $k(\delta)$ if $\kappa = \infty$. If $\kappa < \infty$ then it is easy to see that $k(\delta) \le \kappa$ since

$$\|y^\delta - Tx_\kappa^\delta\| = \|p_\kappa(T)y^\delta\| = \|E_0 y^\delta\| = \|E_0(y^\delta - y)\| \le \delta\,;$$

from this the claim follows because $\tau > 1$.

It should be pointed out that Stopping Rule 3.10 applies to both types of iteration
methods: those of Section 2.1 *and* those of Section 2.3. Note that the residual $r_k = y^\delta - Tx_k^\delta$ is computed anyway in Algorithms 2.1 and 2.3; to obtain its norm, only one
additional inner product is required. Hence, the stopping rule is extremely efficient.

It will be shown next that the iterate determined by Stopping Rule 3.10 has order-
optimal accuracy, i.e., that the error bound (3.17) holds true. In other words, it will be
shown that conjugate gradient algorithms (with parameter $n \ge 1$) are order-optimal
regularization methods, if the iteration is stopped according to the discrepancy principle.

Theorem 3.11 *Let y satisfy Assumption 3.6, and let $\|y - y^\delta\| \le \delta$. If the conjugate
gradient type method (with parameter $n \ge 1$ and right-hand side y^δ) is terminated
after $k(\delta)$ iterative steps according to Stopping Rule 3.10, then*

$$\|T^\dagger y - x_{k(\delta)}^\delta\| \le c\omega^{1/\mu+1}\delta^{\mu/\mu+1} \tag{3.17}$$

with some uniform constant $c > 0$.

Proof. For the sake of notational simplicity, let always be $k = k(\delta)$. As mentioned
above, k is never greater than κ, hence Lemma 3.8 applies at the stopping index. This
yields

$$\|T^\dagger y - x_k^\delta\| \le c(\omega^{1/\mu+1}\delta^{\mu/\mu+1} + |p_k'(0)|\,\delta)\,.$$

To prove the theorem, it remains to show that there is a uniform constant $c > 0$ with

$$|p_k'(0)| \le c\,(\omega/\delta)^{1/\mu+1}\,. \tag{3.18}$$

---

[*]Actually $k(\delta)$ depends on y^δ and on δ, rather than on δ only; however, this dependency is not
important for the subsequent analysis.

46

When $k = 0$ then this is obviously fulfilled, therefore assume $k \geq 1$ in the sequel. First, note that by Lemma 3.7 and Stopping Rule 3.10

$$\tau\delta < \|y^\delta - Tx_{k-1}^\delta\| \leq \delta + c|p_{k-1}'(0)|^{-\mu-1}\omega,$$

where the right-hand side has to be understood as $+\infty$ for $k = 1$. Consequently, (3.18) holds with k replaced by $k - 1$, i.e.,

$$|p_{k-1}'(0)| \leq c\,(\omega/\delta)^{1/\mu+1}. \tag{3.19}$$

The remaining part of the proof is divided in three steps; the idea is to write $|p_k'(0)| = |p_{k-1}'(0)| + |p_k'(0) - p_{k-1}'(0)|$, to use (3.19), and to estimate the expression in Corollary 2.6 to eventually verify (3.18).

Step 1. First, it will be shown that there exists $c > 0$ such that

$$[p_{k-1}, p_{k-1}]_{n-1} \geq c\delta^2 \,(\delta/\omega)^{(n-1)/\mu+1}. \tag{3.20}$$

Note that this is always fulfilled if $n = 1$ by the definition of k (with $c = \tau^2$). For $n > 1$, fix $\varepsilon = (c_\varepsilon\delta/\omega)^{1/\mu+1}$ with a sufficiently small constant $c_\varepsilon > 0$, so that ε is smaller than the smallest root $\lambda_{1,k-1}$ of p_{k-1}; such a choice of c_ε is possible by virtue of (3.19) and (2.19). This in particular implies that

$$0 \leq p_{k-1}(\lambda) \leq 1, \qquad 0 \leq \lambda \leq \varepsilon.$$

For k as defined in Stopping Rule 3.10 follows

$$\begin{aligned}
\tau\delta \ &< \ \|p_{k-1}(T)y^\delta\| \\
&\leq \ \|E_\varepsilon p_{k-1}(T)y^\delta\| + \|(I - E_\varepsilon)p_{k-1}(T)y^\delta\| \\
&\leq \ \|E_\varepsilon y^\delta\| + \varepsilon^{-(n-1)/2}\|(I - E_\varepsilon)p_{k-1}(T)T^{(n-1)/2}y^\delta\| \\
&\leq \ \|E_\varepsilon y^\delta\| + \varepsilon^{-(n-1)/2}[p_{k-1}, p_{k-1}]_{n-1}^{1/2}.
\end{aligned}$$

Using the definition of ε and Assumption 3.6, the first term on the right-hand side can be estimated in terms of δ, namely

$$\|E_\varepsilon y^\delta\| \leq \|E_\varepsilon(y^\delta - y)\| + \|E_\varepsilon T^{\mu+1}w\| \leq \delta + \varepsilon^{\mu+1}\omega = (1 + c_\varepsilon)\delta, \tag{3.21}$$

so that

$$[p_{k-1}, p_{k-1}]_{n-1}^{1/2} \geq (\tau - 1 - c_\varepsilon)\,\delta\,\varepsilon^{(n-1)/2}.$$

Choosing $0 < c_\varepsilon < \tau - 1$, and inserting the definition of ε, assertion (3.20) follows.

Step 2. In the second step, an upper bound will be derived for $[p_{k-1}, p_{k-1}]_{n-1}$; the upper bound depends on the polynomial $p_{k-1}^{[n+1]}$, i.e., the residual polynomial of the conjugate gradient type method with parameter $n+1$. To avoid confusion, the residual polynomials of the algorithm under consideration will therefore be rewritten again by

$\{p_k^{[n]}\}$ in the sequel. The bound to be established below is given in the following inequality, valid for some $c > 0$:

$$[p_{k-1}^{[n]}, p_{k-1}^{[n]}]_{n-1} \le c(\omega/\delta)^{1/\mu+1}[p_{k-1}^{[n+1]}, p_{k-1}^{[n+1]}]_n . \tag{3.22}$$

By Corollary 2.7 $p_{k-1}^{[n+1]}(\lambda)$ is positive in $0 \le \lambda \le \lambda_{1,k-1}^{[n]}$, where – using superscripts again – $\lambda_{1,k-1}^{[n]}$ denotes the smallest root of $p_{k-1}^{[n]}$. Hence, as in Step 1, if ε is chosen to be $\varepsilon = (c_\varepsilon \delta/\omega)^{1/\mu+1}$ with c_ε sufficiently small, then

$$0 \le p_{k-1}^{[n+1]}(\lambda) \le 1, \qquad 0 \le \lambda \le \varepsilon . \tag{3.23}$$

From the optimality property of $p_{k-1}^{[n]}$ therefore follows that

$$
\begin{aligned}
\|T^{(n-1)/2} p_{k-1}^{[n]}(T) y^\delta\| \quad &\le \quad \|T^{(n-1)/2} p_{k-1}^{[n+1]}(T) y^\delta\| \\
&\le \quad \|E_\varepsilon p_{k-1}^{[n+1]}(T) T^{(n-1)/2} y^\delta\| + \|(I - E_\varepsilon) p_{k-1}^{[n+1]}(T) T^{(n-1)/2} y^\delta\| \\
&\le \quad \|E_\varepsilon T^{(n-1)/2} y^\delta\| + \varepsilon^{-1/2} \|(I - E_\varepsilon) p_{k-1}^{[n+1]}(T) T^{n/2} y^\delta\| \\
&\le \quad \varepsilon^{(n-1)/2} \|E_\varepsilon y^\delta\| + \varepsilon^{-1/2} \|p_{k-1}^{[n+1]}(T) T^{n/2} y^\delta\| .
\end{aligned}
$$

$\|E_\varepsilon y^\delta\|$ can be estimated as in (3.21), hence

$$[p_{k-1}^{[n]}, p_{k-1}^{[n]}]_{n-1}^{1/2} \le (1 + c_\varepsilon) \varepsilon^{(n-1)/2} \delta + \varepsilon^{-1/2} [p_{k-1}^{[n+1]}, p_{k-1}^{[n+1]}]_n^{1/2} . \tag{3.24}$$

For $n = 1$ the stopping rule definition of k allows to estimate δ from above by $[p_{k-1}^{[1]}, p_{k-1}^{[1]}]_0^{1/2}/\tau$; inserting this into (3.24) one obtains

$$(1 - \frac{1 + c_\varepsilon}{\tau})^2 [p_{k-1}^{[1]}, p_{k-1}^{[1]}]_0 \le \varepsilon^{-1} [p_{k-1}^{[2]}, p_{k-1}^{[2]}]_1 .$$

Thus, choosing $0 < c_\varepsilon < \tau - 1$ and inserting the definition of ε, (3.22) follows for $n = 1$. If $n > 1$ then (3.20) can be used to estimate $\varepsilon^{(n-1)/2} \delta$ in (3.24) from above, hence

$$[p_{k-1}^{[n]}, p_{k-1}^{[n]}]_{n-1} \le 2(1 + c_\varepsilon)^2 c_\varepsilon^{(n-1)/\mu+1} c^{-1} [p_{k-1}^{[n]}, p_{k-1}^{[n]}]_{n-1} + 2\varepsilon^{-1} [p_{k-1}^{[n+1]}, p_{k-1}^{[n+1]}]_n .$$

As $n > 1$, the factor in front of the first term on the right-hand side can be made smaller than $1/2$, say, by choosing c_ε sufficiently small; combining terms, this leads to

$$\frac{1}{2} [p_{k-1}^{[n]}, p_{k-1}^{[n]}]_{n-1} \le 2\varepsilon^{-1} [p_{k-1}^{[n+1]}, p_{k-1}^{[n+1]}]_n ,$$

and (3.22) follows from the definition of ε.

48

Final Step. Using (3.22), Corollary 2.6 yields

$$0 \le p_{k-1}^{[n]}{}'(0) - p_k^{[n]\prime}(0) \le \frac{[p_{k-1}^{[n]}, p_{k-1}^{[n]}]_{n-1}}{[p_{k-1}^{[n+1]}, p_{k-1}^{[n+1]}]_n} \le c(\omega/\delta)^{1/\mu+1}.$$

Together with (3.19) and the triangle inequality, (3.18) follows and the proof of the theorem is complete. $\qquad\square$

Remark. It is obvious from the remark following Corollary 3.9, that in order to prove the same result for the conjugate gradient type methods of Section 2.3 one has to show

$$|p_k'(0)| \le c\,(\omega/\delta)^{2/\mu+1}$$

instead of (3.18). This is easily established with the same arguments as above when choosing $\varepsilon = (c_\varepsilon \delta/\omega)^{2/\mu+1}$ instead.

Although Theorem 3.11 only considers data satisfying Assumption 3.6, it can nevertheless be used to conclude that Stopping Rule 3.10 leads to converging approximations of $T^\dagger y$, regardless whether Assumption 3.6 holds or not.

Theorem 3.12 *Let* $y \in \mathcal{R}(T)$ *and* $\|y - y^\delta\| \le \delta$. *If the stopping index* $k(\delta)$ *for a conjugate gradient type method with parameter* $n \ge 1$ *is determined according to Stopping Rule 3.10 then* $x_{k(\delta)}^\delta \to T^\dagger y$ *as* $\delta \to 0$.

Proof. The idea of the proof is to consider y^δ as a perturbation of $(I - E_\varepsilon)y$ with a clever choice of $\varepsilon = \varepsilon(\delta)$. Let $\tau > 1$ be the parameter in Stopping Rule 3.10: with this, define

$$\varepsilon(\delta) := \frac{1}{2} \inf \{\varepsilon > 0 \,|\, \|E_\varepsilon y\| \ge \frac{\tau - 1}{2}\delta\}, \qquad \delta > 0.$$

Since $y \in \mathcal{R}(T)$, $E_\varepsilon y \to 0$ as $\varepsilon \to 0$, hence $\varepsilon(\delta)$ is strictly positive and the additional factor $1/2$ in front of the *inf* guarantees that $\|E_{\varepsilon(\delta)}y\| \le \frac{1}{2}(\tau - 1)\delta$. In the sequel, the argument of ε is omitted for notational convenience. It follows that

$$\|(I - E_\varepsilon)y - y^\delta\| \le \|y - y^\delta\| + \|E_\varepsilon y\| \le \frac{\tau + 1}{2}\delta =: \tilde{\delta}.$$

Introducing $\tilde{\tau} = 2\tau/(\tau + 1) > 1$, one has $\tilde{\tau}\tilde{\delta} = \tau\delta$, and hence the discrepancy principle returns precisely the same stopping index $\tilde{k}(\tilde{\delta}) = k(\delta)$, when applied with parameter $\tilde{\tau}$ instead of τ and data error bound $\tilde{\delta}$ instead of δ.

Note that $T^\dagger(I - E_\varepsilon)y \in \mathcal{R}(T^\mu)$ for every $\mu > 0$, namely $T^\dagger(I - E_\varepsilon)y = T^\mu w_\varepsilon$ with

$$\|w_\varepsilon\|^2 = \int_{\varepsilon+}^\infty \lambda^{-2\mu-2}d\|E_\lambda y\|^2 \le \varepsilon^{-2\mu}\int_{\varepsilon+}^\infty \lambda^{-2}d\|E_\lambda y\|^2 \le \varepsilon^{-2\mu}\|T^\dagger y\|^2.$$

It therefore follows from Theorem 3.11 that

$$\|T^\dagger(I - E_\varepsilon)y - x_{k(\delta)}^\delta\| \leq \tilde{c}\|w_\varepsilon\|^{1/\mu+1}\tilde{\delta}^{\mu/\mu+1} \leq c(\delta/\varepsilon)^{\mu/\mu+1}\|T^\dagger y\|^{1/\mu+1},$$

and consequently,

$$\|T^\dagger y - x_{k(\delta)}^\delta\| \leq \|E_\varepsilon T^\dagger y\| + c(\delta/\varepsilon)^{\mu/\mu+1}\|T^\dagger y\|^{1/\mu+1}.$$

It remains to show that $\delta/\varepsilon \to 0$ and $E_\varepsilon T^\dagger y \to 0$ as $\delta \to 0$. If $\lim \varepsilon(\delta) = \varepsilon_0 > 0$ as $\delta \to 0$ then this is obvious, since

$$\|E_{\varepsilon_0}y\| \leq \|E_{\varepsilon(\delta)}y\| \leq \frac{1}{2}(\tau - 1)\delta$$

for every $\delta > 0$. This implies

$$0 = \|E_{\varepsilon_0}y\| = \|E_{\varepsilon_0}T^\dagger y\| = \lim_{\delta \to 0}\|E_{\varepsilon(\delta)}T^\dagger y\|.$$

On the other hand, if $\varepsilon \to 0$ as $\delta \to 0$ then, by definition of $\varepsilon = \varepsilon(\delta)$ and the continuity of E_λ from the right,

$$\frac{\tau - 1}{2}\delta \leq \|E_{2\varepsilon}y\| \leq 2\varepsilon\|E_{2\varepsilon}T^\dagger y\|,$$

hence, $\delta/\varepsilon \leq 4(\tau - 1)^{-1}\|E_{2\varepsilon}T^\dagger y\|$. Since $\varepsilon \to 0$, the assertion follows. $\square$

Remark. For conjugate gradient type methods with TT^* instead of T one proceeds similarly. Since in this case $\|w_\varepsilon\| = O(\varepsilon^{-\mu/2})$ one eventually has to show that $\delta^2/\varepsilon \to 0$ as $\delta \to 0$, which follows from the same argument as above.

Note that the "approximate preimage" $T^\dagger(I - E_\varepsilon)y$ occurring in the proof of Theorem 3.12 is nothing else than the familiar truncated spectral expansion approximation mentioned in (2.15).

3.4 A heuristic stopping rule

While the discrepancy principle provides an efficient (and simple) stopping rule when an upper bound δ for the perturbation $\|y - y^\delta\|$ is available, difficulties arise when there is only incomplete information about the magnitude of the perturbation. In fact, in many practical applications this is the case. This justifies the increasing interest in so-called *heuristic stopping rules* that avoid any knowledge of δ as opposed to order-optimal stopping rules which require knowledge of δ to guarantee order-optimal error bounds. With a heuristic stopping rule the stopping index k depends solely on information gathered in the course of the iteration, and therefore only on the right-hand side y^δ.

One possibility for constructing a heuristic stopping rule originates from a comparison of the error estimate (3.16) of Corollary 3.9 with the inequality (3.6) of Lemma 3.7.

The idea is to consider these inequalities as equalities, motivated by the fact that both bounds are reasonably sharp. This suggests the following approximation, which should roughly hold up to a multiplicative constant, namely

$$\|T^\dagger y - x_k^\delta\| \approx |p_k'(0)| \, \|y^\delta - T x_k^\delta\| \, ,$$

and respectively for the algorithms of Section 2.3 using TT^*:

$$\|T^\dagger y - x_k^\delta\| \approx |p_k'(0)|^{1/2} \|y^\delta - T x_k^\delta\| \, .$$

The right-hand sides can easily be computed in the course of the iteration, cf. recursion (2.11) for $|p_k'(0)|$.

Note that the residual norm factor in these error estimates decreases towards the noise level as $k \to \infty$ (cf. Lemma 3.7) whereas the other factor tends to infinity. Therefore, for large values of k, $|p_k'(0)|\delta$ and $|p_k'(0)|^{1/2}\delta$, respectively, will be reasonable approximations of the right-hand sides, showing that these estimates typically diverge as $k \to \infty$. This is the basis for the following stopping rule.

Stopping Rule 3.13 *For conjugate gradient type methods as considered in Section 2.1 compute*

$$\eta_0 = \|y^\delta\|, \qquad \eta_k = |p_k'(0)| \, \|y^\delta - T x_k^\delta\|, \quad k \geq 1; \qquad (3.25)$$

for the methods of Section 2.3 compute instead

$$\eta_0 = \|y^\delta\|, \qquad \eta_k = |p_k'(0)|^{1/2} \|y^\delta - T x_k^\delta\|, \quad k \geq 1.$$

In either case terminate the iteration after $k(y^\delta)$ steps, provided $\eta_{k(y^\delta)} \leq \eta_k$ for all $0 \leq k \leq \kappa$.

In spite of the aforementioned motivation it can occur that the sequence $\{\eta_k\}$ has no global minimum at all, i.e., $\eta_k \to 0$ as $k \to \infty$. In this case the above stopping rule fails. For actual computations one anyhow has to relax the stipulation that $\eta_{k(y^\delta)}$ is really the global minimum. Instead, one will typically perform a few look-ahead steps to achieve certainty whether a local minimum of $\{\eta_k\}$ is likely to be the global minimum. Finally, the global minimum may be zero, i.e., $\|y^\delta - T x_{k(y^\delta)}^\delta\|$ vanishes; in this case, $x_{k(y^\delta)}^\delta = T^\dagger y^\delta$ and it depends on the perturbation y^δ whether this is a good approximation or not (compare Section 2.6). Otherwise, the following error bound can be derived.

Theorem 3.14 *Let y satisfy Assumption 3.6 and let $\|y - y^\delta\| \le \|y\|$. If Stopping Rule 3.13 determines a finite stopping index $k(y^\delta)$, and if $\delta_* = \|y^\delta - Tx^\delta_{k(y^\delta)}\| \neq 0$ then*

$$\|T^\dagger y - x^\delta_{k(y^\delta)}\| \le c(1 + \frac{\|y - y^\delta\|}{\delta_*})\omega^{1/\mu+1}\rho_*^{\mu/\mu+1}, \qquad (3.26)$$

where $\rho_* = \max\{\|y - y^\delta\|, \delta_*\}$.

Proof. Assume without loss of generality that $\|y - y^\delta\| = \delta$. Since $p'_0 \equiv 0$, (3.25) immediately implies

$$|p'_k(0)|\,\|y^\delta - Tx^\delta_k\| \le \eta_k \qquad \text{for every } k \in \mathbf{N}_0.$$

Hence Lemma 3.8 yields

$$\|T^\dagger y - x^\delta_{k(y^\delta)}\| \le c\,(\omega^{1/\mu+1}\rho_*^{\mu/\mu+1} + |p'_{k(y^\delta)}(0)|\,\delta) \le c\,(\omega^{1/\mu+1}\rho_*^{\mu/\mu+1} + \frac{\delta}{\delta_*}\eta_{k(y^\delta)}), \quad (3.27)$$

and it remains to estimate $\eta_{k(y^\delta)}$. By definition, $\eta_{k(y^\delta)} \le \eta_{k(\delta)}$, where $k(\delta)$ denotes the iteration index chosen by Stopping Rule 3.10. Two cases must be considered: if $k(\delta) = 0$ then $\|y^\delta\| \le \tau\delta$, and it follows readily from the given assumption $\delta \le \|y\|$ that

$$\eta_{k(y^\delta)} \le \|y^\delta\| \le \tau\delta \le \tau\|y\|^{1/\mu+1}\delta^{\mu/\mu+1} \le \tau\omega^{1/\mu+1}\delta^{\mu/\mu+1}; \qquad (3.28)$$

if $k(\delta) \neq 0$ then (3.18) yields the same bound:

$$\eta_{k(y^\delta)} \le |p'_{k(\delta)}(0)|\,\|y^\delta - Tx^\delta_{k(\delta)}\| \le c\tau\,(\omega/\delta)^{1/\mu+1}\delta.$$

Inserting this into (3.27) completes the proof. $\qquad\qquad\square$

In principle, this result allows an a posteriori justification of the stopping index $k(y^\delta)$ by computing δ_*: if δ_* has about the expected order of the noise level then Theorem 3.14 shows that the error has order-optimal accuracy. On the other hand, if δ_* is seemingly too large, then the right-hand side of (3.26) will stagnate. More critical situations arise when δ_* is much too small, i.e., when $k(y^\delta)$ is too large. In this case $\|y - y^\delta\|/\delta_*$ will blow up and the approximations may diverge.

Of particular interest is therefore the following case study. Assume that the closure of the range of T is a proper subset of $\mathcal{X}$, and that the perturbation $y - y^\delta$ has a nontrivial component along the orthogonal complement of $\mathcal{R}(T)$, i.e.,

$$\|(I - P)(y - y^\delta)\| \ge \gamma\|y - y^\delta\| \neq 0 \qquad (3.29)$$

for some $\gamma > 0$. Since $y \in \mathcal{R}(T)$ this obviously implies

$$\|y^\delta - Tx^\delta_k\| \ge \|(I - P)(y^\delta - Tx^\delta_k)\| = \|(I - P)y^\delta\| \ge \gamma\|y - y^\delta\|. \qquad (3.30)$$

This guarantees that the stopping index $k(y^\delta)$ of Stopping Rule 3.13 is a well-defined finite number, and δ_* as defined in Theorem 3.14 is always greater than $\gamma\|y - y^\delta\|$. Therefore, the following corollary is valid:

Corollary 3.15 *If, in addition to the assumptions of Theorem 3.14, (3.29) holds for some $\gamma > 0$, then the iterate $x^\delta_{k(y^\delta)}$ resulting from Stopping Rule 3.13 satisfies*

$$\|T^\dagger y - x^\delta_{k(y^\delta)}\| \le \frac{c}{\gamma}\omega^{1/\mu+1}\rho_*^{\mu/\mu+1},$$

where ρ_* is as above.

Even more can be said, and it must be pointed out that the following result does not require that Assumption 3.6 holds:

Theorem 3.16 *Let $y \in \mathcal{R}(T)$ and let $\{y^\delta\}_{\delta > 0}$ be a sequence of perturbations of y with $y^\delta \to y$ as $\delta \to 0$ in such a way that (3.29) holds uniformly for some $\gamma > 0$. If $k(y^\delta)$ is determined by Stopping Rule 3.13, then the iterates $x^\delta_{k(y^\delta)}$ converge to $T^\dagger y$ as $\delta \to 0$.*

Proof. Consider first the case $y = 0$. The definition of $\eta_{k(y^\delta)}$ and (3.30) yield

$$\|y - y^\delta\| = \eta_0 \ge \eta_{k(y^\delta)} \ge \gamma\max\{1, |p'_{k(y^\delta)}(0)|\}\,\|y - y^\delta\|.$$

It follows that $|p'_{k(y^\delta)}(0)| \le 1/\gamma$, independent of y^δ, hence (3.16) yields

$$\|T^\dagger y - x^\delta_{k(y^\delta)}\| \le c\,|p'_{k(y^\delta)}(0)|\,\|y - y^\delta\| \le \frac{c}{\gamma}\|y - y^\delta\|.$$

Thus, if $y = 0$ then $x^\delta_{k(y^\delta)} \to T^\dagger y$ as $\delta \to 0$.

The proof for $y \ne 0$ will be given in two steps: first, it will be shown that $\rho_* \to 0$ as $\delta \to 0$; then, convergence follows eventually with the help of (3.26). In both steps, the key argument is an approximation of y by elements in $\mathcal{R}(T^{\mu+1})$ like in the proof of Theorem 3.12, namely by $(I - E_\varepsilon)y = T^{\mu+1}w_\varepsilon, \varepsilon > 0$. Here, μ is some arbitrary positive number that will remain fixed throughout the proof. Recall that for $\rho \ge 2\|y - y^\delta\|$ (note that $y \ne y^\delta$ by virtue of (3.29)), $\varepsilon = \varepsilon(\rho)$ can be constructed as in the proof of Theorem 3.12 in such a way that

$$\|(I - E_\varepsilon)y - y^\delta\| \le \rho, \qquad (3.31)$$

and, when $\delta \to 0$ and $\rho \to 0$ subject to $\rho \ge 2\|y - y^\delta\|$, one has

$$E_\varepsilon T^\dagger y \to 0 \quad\text{and}\quad \|w_\varepsilon\|^{1/\mu+1}\rho^{\mu/\mu+1} \to 0. \qquad (3.32)$$

Step 1. Let ρ_* be as in Theorem 3.14; the aim of this first step is to show that $\rho_* \to 0$ as $\delta \to 0$. For $\delta > 0$ and right-hand side y^δ, construct $\varepsilon = \varepsilon(\rho)$ as above with

$$\rho = 2\|y - y^\delta\|.$$

Consider the iteration index $k(\rho)$ as determined by the discrepancy principle (with some fixed $\tau > 1$). $(I - E_\epsilon)y$ satisfies the assumptions of Theorem 3.11 by virtue of (3.31). Therefore, cf. (3.18),

$$\|y^\delta - Tx^\delta_{k(\rho)}\| \le 2\tau \|y - y^\delta\|, \qquad |p'_{k(\rho)}(0)| \le c(\|w_\epsilon\|/\rho)^{1/\mu+1}.$$

Consequently,

$$\eta_{k(\rho)} = \begin{cases} \|y^\delta - Tx^\delta_{k(\rho)}\| \le 2\tau \|y - y^\delta\|, & k(\rho) = 0, \\ |p'_{k(\rho)}(0)| \, \|y^\delta - Tx^\delta_{k(\rho)}\| \le c\tau \|w_\epsilon\|^{1/\mu+1} \rho^{\mu/\mu+1}, & k(\rho) \ne 0. \end{cases}$$

By (3.32) the right-hand side goes to zero as $\delta \to 0$, showing

$$\eta_{k(y^\delta)} \le \eta_{k(\rho)} \to 0, \qquad \delta \to 0.$$

Since $\delta_* = \|y^\delta - Tx^\delta_{k(y^\delta)}\| \le \eta_{k(y^\delta)}$ by virtue of (3.25), $\delta_* \to 0$ as $\delta \to 0$ and therefore $\rho_* \to 0$ as $\delta \to 0$, which was to be shown.

Step 2. In this second step, let $\epsilon = \epsilon(2\rho_*)$ be such that (3.31) and (3.32) hold with $\rho = 2\rho_*$; recall that $2\rho_*$ is greater than $2\|y - y^\delta\|$ by definition of ρ_*. Since it has been shown in Step 1 that $\rho_* \to 0$ as $\delta \to 0$, it follows from (3.32) that $E_\epsilon y \to 0$ as $\delta \to 0$. Since $y \ne 0$ in this part of the proof,

$$\|(I - E_\epsilon)y - y^\delta\| \le \|y - y^\delta\| + \|E_\epsilon y\| < \|(I - E_\epsilon)y\|$$

for δ sufficiently small, showing that $(I - E_\epsilon)y$ satisfies the assumptions on y of Theorem 3.14. Hence, (3.26) yields

$$\|T^\dagger(I - E_\epsilon)y - x^\delta_{k(y^\delta)}\| \le c\left(1 + \frac{\|(I - E_\epsilon)y - y^\delta\|}{\delta_*}\right)\|w_\epsilon\|^{1/\mu+1} \rho^{\mu/\mu+1}_{**}, \qquad (3.33)$$

where, cf. (3.31),

$$\rho_{**} = \max\{\|(I - E_\epsilon)y - y^\delta\|, \delta_*\} \le 2\rho_*.$$

By virtue of (3.30) $\|y - y^\delta\| \le \delta_*/\gamma$, hence (3.31) implies

$$\|(I - E_\epsilon)y - y^\delta\| \le 2\rho_* = 2\max\{\|y - y^\delta\|, \delta_*\} \le \frac{2}{\gamma}\delta_*.$$

Inserting the previous two estimates into (3.33), one obtains

$$\|T^\dagger(I - E_\epsilon)y - x^\delta_{k(y^\delta)}\| \le c\|w_\epsilon\|^{1/\mu+1} \rho^{\mu/\mu+1}_*,$$

and it follows that

$$\begin{aligned} \|T^\dagger y - x^\delta_{k(y^\delta)}\| &\le \|T^\dagger y - T^\dagger(I - E_\epsilon)y\| + \|T^\dagger(I - E_\epsilon)y - x^\delta_{k(y^\delta)}\| \\ &\le \|E_\epsilon T^\dagger y\| + c\|w_\epsilon\|^{1/\mu+1} \rho^{\mu/\mu+1}_*. \end{aligned}$$

Since $\rho_* \to 0$ as $\delta \to 0$, the right-hand side converges to zero by virtue of (3.32), which completes the proof. $\qquad \square$

The modifications of the proof for the algorithms of Section 2.3 are straightforward.

Notes and remarks

Section 3.1. The observation that the iteration error for the classical conjugate gradient method decreases monotonically can already be found in HESTENES and STIEFEL [45]. The idea to use Lemma 3.1 (due to ASKEY [3]) for proving Theorem 3.2 goes back to BRAKHAGE [6]. Both results can be extended to noninteger values of m and n, cf. TRENCH [80].

GILYAZOV [22] (see also [24, p. 40]) was probably the first to come up with a proof of Theorem 3.4 for the important case $n = 1$. It is his proof (rewritten in terms of orthogonal polynomials) that is presented here. A different proof using arguments similar to those of Section 3.2 was subsequently given by NEMIROVSKII and POLYAK [59]. Partial convergence results have been obtained earlier by KAMMERER and NASHED [46]. LARDY [51] extended Theorem 3.4 to all $n > 1$.

Section 3.2. The exposition of this section follows NEMIROVSKII [58]; see also PLATO [68]. Those authors only treat the case $n = 1$, but also consider perturbations (resp. discretizations) of the operator T.

Section 3.3. The order-optimality of the discrepancy principle for MR and CGNE has been proven by NEMIROVSKII [58] with a different argument. It should be remarked that GILYAZOV [23] obtained suboptimal convergence rates for a modification of the discrepancy principle at about the same time. ALIFANOV and RUMJANCEV [1] established earlier the regularizing properties (but not the order-optimality) of a somewhat different stopping rule for CGNE.

Theorem 3.12 is due to PLATO [67].

Section 3.4. Stopping Rule 3.13 has been suggested in [40]. The computation of $|p_k'(0)|$ for diagnostic purposes has first been proposed by LOUIS [54]. Concerning heuristic stopping rules in general, BAKUSHINSKII [4] has shown that one can never design a regularizing algorithm with no prior knowledge beyond the given data y^δ. Nevertheless, as can be seen from Theorem 3.14, regularizing properties can be restored under certain conditions. An earlier result with a similar flavour (but in connection with a heuristic parameter choice rule for Tikhonov regularization) was proven by LEONOV [52].

Other heuristic stopping rules for CGNE have been considered by HANSEN [42]. No error bounds are known for these methods.

4. Regularizing Properties of CG and CGME

The analysis of CG and CGME leads to new difficulties, and some of the main results turn out to be different from those of the previous chapter. In the first section of the present chapter it will be shown that CG and CGME diverge whenever $y \notin \mathcal{R}(T)$; recall from the remark following Theorem 3.4 that CGNE converges to $T^\dagger y$ for all $y \in \mathcal{D}(T^\dagger)$. Other differences concern the behavior of the residuals. In Example 4.3 it will be shown that the norm of the CG residuals does not decrease monotonically. More striking, the discrepancy principle (which is based on the residual norm) does *not* regularize CG (CGME): a counterexample is constructed in Section 4.2. A different stopping rule is easily implemented, though, providing order-optimal accuracy for the resulting approximation. There is also a heuristic stopping rule in analogy to the rule in Section 3.4.

4.1 Monotonicity, convergence and divergence

The first two results of this section establish very similar convergence results for CG and CGME as have been obtained for the other conjugate gradient type methods in Section 3.1.

Theorem 4.1 *If $y \in \mathcal{R}(T)$ and $\{x_k\}$ are the iterates of CG or CGME then $\|T^\dagger y - x_k\|$ is strictly decreasing for $0 \leq k \leq \kappa$.*

Proof. Since $T^\dagger y - x_k = p_k(T)T^\dagger y$ one has $\|T^\dagger y - x_k\|^2 = [p_k, p_k]_{-2}$ for CG and $\|T^\dagger y - x_k\|^2 = [p_k, p_k]_{-1}$ for CGME, respectively. Here, $p_k = p_k^{[0]}$, and hence the assertion follows from Theorem 3.2. $\qquad\square$

As in the previous chapter the convergence of CG (CGME) follows from the monotonicity of the iteration error.

Theorem 4.2 *If $y \in \mathcal{R}(T)$ then the iterates of CG (CGME) converge to $T^\dagger y$.*

Proof. Let $y \in \mathcal{R}(T)$ and $x = T^\dagger y$. If the iteration terminates for some finite κ then the statement of the theorem is true; see Section 2.1. Therefore, assume that the iteration does not terminate. In the case of CGME Proposition 2.1 holds accordingly for $n = 0$, i.e., if $\varphi_k \in \Pi_k^0$ then

$$\|T^\dagger y - x_k\|^2 = [p_k, p_k]_{-1} \leq [\varphi_k, \varphi_k]_{-1}.$$

Choosing $\varphi_k(\lambda) = (1-\lambda)^k$ one obtains $[\varphi_k, \varphi_k]_{-1} = \|(I - T^*T)^k x\|^2$, which converges to zero as $k \to \infty$ by the Banach-Steinhaus theorem. The assertion is therefore true for CGME. The proof for CG is the same as the one of Theorem 3.4. $\qquad\square$

There are well-known examples, however, which show that the residual norm need not decrease monotonically during the CG (CGME) iteration. This is different to the other conjugate gradient type methods, cf. Corollary 3.3 (i). One such example is the following.

Example 4.3 Let T be selfadjoint positive definite with eigenfunctions v_1 and v_2, and

$$Tv_1 = \tau v_1, \quad Tv_2 = v_2, \quad 0 < \tau < 1.$$

Choosing $y = \eta v_1 + v_2$ one has for $p(\lambda) = 1 - \gamma\lambda \in \Pi_1^0$:

$$[p, p]_0 = \eta^2(1 - \gamma\tau)^2 + (1 - \gamma)^2, \qquad [p, p]_{-1} = \frac{\eta^2}{\tau}(1 - \gamma\tau)^2 + (1 - \gamma)^2.$$

From the fact that the residual polynomial p_1 of CG minimizes $[p, p]_{-1}$ among all $p \in \Pi_1^0$ one finds $\gamma = (1 + \eta^2)(1 + \eta^2\tau)^{-1}$, i.e.,

$$p_1(\lambda) = 1 - \frac{1 + \eta^2}{1 + \eta^2\tau}\lambda.$$

Consequently, one has

$$\|y - Tx_1\|^2 = [p_1, p_1]_0 = \eta^2(1 - \frac{1 + \eta^2}{1 + \eta^2\tau}\tau)^2 + (1 - \frac{1 + \eta^2}{1 + \eta^2\tau})^2 = \eta^2(1 + \eta^2)(\frac{1 - \tau}{1 + \eta^2\tau})^2,$$

whereas $\|y - Tx_0\|^2 = \|y\|^2 = 1 + \eta^2$. Thus, when τ is sufficiently close to zero and $\eta > 1$ it follows that

$$\|y - Tx_0\| < \|y - Tx_1\| \approx \eta\|y - Tx_0\|.$$

Finally, consider the following divergence result, and its difference from the corresponding Theorem 3.5 for conjugate gradient type methods with parameter $n \geq 1$.

Theorem 4.4 *If* $y \notin \mathcal{R}(T)$ *and* $\{x_k\}$ *are the iterates of* CG (CGME) *then either* $\kappa < \infty$ *and the iteration breaks down in the* $\kappa + 1$st *step with division by zero, or* $\kappa = \infty$ *and* $\|x_k\| \to \infty$ *as* $k \to \infty$.

Proof. Assume first that $Py \neq y$, i.e., $E_0 y \neq 0$. If $\|E_\lambda y\|^2$ has just a finite number κ of positive points of increase, then the iteration breaks down in the $\kappa + 1$st step, see

58

Section 2.1. Otherwise, by virtue of (2.19), $|p_k'(0)| \geq k$, showing that the CG-iterates x_k diverge to infinity in norm:

$$\|x_k\| \geq q_{k-1}(0)\|E_0 y\| = |p_k'(0)|\,\|E_0 y\| \longrightarrow \infty, \qquad k \to \infty.$$

For CGME a different argument is required, since $x_k = T^* q_{k-1}(TT^*)y$ has no component in $\mathcal{N}(T)$. Expanding q_{k-1} in the basis $\{p_j^{[1]}\}$ gives

$$q_{k-1} = \sum_{j=0}^{k-1} \frac{[q_{k-1}, p_j^{[1]}]_1}{[p_j^{[1]}, p_j^{[1]}]_1}\, p_j^{[1]},$$

and hence,

$$\|x_k\|^2 = [q_{k-1}, q_{k-1}]_1 = \sum_{j=0}^{k-1} \frac{[q_{k-1}, p_j^{[1]}]_1^2}{[p_j^{[1]}, p_j^{[1]}]_1}. \tag{4.1}$$

Since $q_{k-1} = (1 - p_k)/\lambda$, cf. (2.2), (2.23) implies that

$$[q_{k-1}, p_j^{[1]}]_1 = [1, p_j^{[1]}]_0 - [p_k^{[0]}, p_j^{[1]}]_0 = [1, p_j^{[1]}]_0 = [p_j^{[1]}, p_j^{[1]}]_0, \qquad 0 \leq j < k,$$

where the right-hand side can be estimated from below by

$$[p_j^{[1]}, p_j^{[1]}]_0 = \|p_j^{[1]}(TT^*)y\|^2 \geq \|E_0 y\|^2, \qquad 0 \leq j < k.$$

Inserting this into (4.1), Lemma 2.4 yields, cf. (2.21),

$$\|x_k\|^2 \geq \sum_{j=0}^{k-1} \frac{\|E_0 y\|^4}{[p_j^{[1]}, p_j^{[1]}]_1} = \|E_0 y\|^4\,[p_{k-1}^{[2]}, p_{k-1}^{[2]}]_1^{-1}.$$

By Proposition 2.1 $[p_{k-1}^{[2]}, p_{k-1}^{[2]}]_1$ tends to zero as $k \to \infty$, and hence, $\|x_k\| \to \infty$ as was to be shown.

Assume next that y belongs to the closure of $\mathcal{R}(T)$. In this case the orthogonality of $\{p_k^{[0]}\}$ implies that $[q_{k-1}, p_j^{[1]}]_1 = [1/\lambda, p_j^{[1]}]_1$, $0 \leq j \leq k - 1$, and hence (4.1) can be rewritten as

$$[q_{k-1}, q_{k-1}]_1 = \sum_{j=0}^{k-1} \frac{[1/\lambda, p_j^{[1]}]_1^2}{[p_j^{[1]}, p_j^{[1]}]_1}. \tag{4.2}$$

Note that the terms on the right-hand side are independent of k, and therefore the numbers $[q_{k-1}, q_{k-1}]_1$ are nondecreasing (as function of k). Assume next that the sequence (4.2) remains bounded: this implies that the formal series expansion

$$1/\lambda \sim \sum_{j=0}^{\infty} \frac{[1/\lambda, p_j^{[1]}]_1}{[p_j^{[1]}, p_j^{[1]}]_1}\, p_j^{[1]}$$

defines a function $1/\lambda$ in the topology induced by the norm $[\cdot, \cdot]_1$, i.e.,

$$\int_0^\infty \frac{1}{\lambda} \, d\|E_\lambda y\|^2 < \infty. \tag{4.3}$$

For CGME, $[q_{k-1}, q_{k-1}]_1 = \|x_k\|^2$ and (4.3) is equivalent to $y \in \mathcal{R}(T)$, showing that bounded subsequences of $\{x_k\}$ can only exist if $y \in \mathcal{R}(T)$. Thus, the theorem is proved for CGME. For CG the argument is as follows: if some subsequence of $\{x_k\}$ remains bounded then $\|T^{1/2}x_k\|^2 = [q_{k-1}, q_{k-1}]_1 \le c < \infty$ for all $k \in \mathbb{N}$, and (4.3) implies that $y \in \mathcal{R}(T^{1/2})$, i.e., there exists $z \in \mathcal{X}$ with $T^{1/2}z = y$. Furthermore, $T^{1/2}x_k \to z$ as $k \to \infty$, since $q_{k-1} \to 1/\lambda$ in the above topology. It follows that $Tx_k \to T^{1/2}z = y$. On the other hand, when $\{x_k\}$ is bounded then there is a subsequence of $\{x_k\}$ which converges weakly to $x \in \mathcal{X}$, say. Since the corresponding images $\{Tx_k\}$ converge weakly to Tx, one has $y = Tx \in \mathcal{R}(T)$, which gives the desired contradiction. □

4.2 Failure of the discrepancy principle: a counterexample

Let $d\alpha(t)$ be the Lebesgue measure on $[0, 1]$ with an additional unit mass placed at $t = 0$. Denote by $\mathcal{X} = \mathcal{L}^2(d\alpha)$ the space of functions x over $[0, 1]$ which are square integrable with respect to $d\alpha$. Thus, for $x, z \in \mathcal{X}$ the inner product $\langle \cdot, \cdot \rangle$ is defined by

$$\langle x, z \rangle^2 = x(0)z(0) + \int_0^1 x(t)z(t) \, dt. \tag{4.4}$$

Define $T : \mathcal{X} \to \mathcal{X}$ by

$$(Tx)(t) = tx(t), \qquad x \in \mathcal{X}, \, t \in [0, 1]. \tag{4.5}$$

T is selfadjoint, semidefinite, with $\|T\| = 1$ and nullspace

$$\mathcal{N}(T) = \{x \in \mathcal{X} \mid x = 0 \text{ a.e. in } (0, 1]\}.$$

Note that this is a nontrivial nullspace due to the definition of $\mathcal{X}$, cf. (4.4). The orthoprojectors of the spectral family $\{E_\lambda\}$ of T are given by $E_\lambda \equiv 0$ for $\lambda < 0$, $E_\lambda \equiv I$ for $\lambda \ge 1$, and

$$(E_\lambda x)(t) = \begin{cases} x(t), & 0 \le t \le \lambda, \\ 0, & \lambda < t \le 1, \end{cases} \qquad 0 \le \lambda < 1.$$

Consequently, for $0 \le \lambda \le 1$,

$$\|E_\lambda x\|^2 = x^2(0) + \int_0^\lambda x^2(t) \, dt, \qquad d\|E_\lambda x\|^2 = x^2(\lambda)d\lambda.$$

For $\mu > 0$ one readily obtains $(T^\mu x)(t) = t^\mu x(t)$, hence the elements of $\mathcal{R}(T^\mu)$ can be described by their behavior near $t = 0$. For example, if $\nu > 0$ then

60

$$x^{(\nu)}(t) := \begin{cases} 0, & t = 0, \\ t^{\nu-1/2}(1-t)^{-1/4}, & 0 < t < 1, \end{cases} \tag{4.6}$$

belongs to $\mathcal{R}(T^\mu)$ for every $\mu < \nu$.

Let $y := Tx^{(\nu)}$, and consider the solution of the equation

$$Tx = y. \tag{4.7}$$

CG applied to this problem will generate residual polynomials $\{p_k^{[0]}\}$ that are orthogonal with respect to the inner product

$$[\varphi, \psi]_0 = \int_0^1 \varphi(\lambda)\psi(\lambda)\, \lambda^{2\nu+1}(1-\lambda)^{-1/2}\, d\lambda. \tag{4.8}$$

This means that the residual polynomials are translated and rescaled Jacobi polynomials, i.e.,

$$p_k^{[0]}(\lambda) = P_k^{(2\nu+1,-1/2)}(1-2\lambda)/P_k^{(2\nu+1,-1/2)}(0).$$

In the remainder of this section the inner product $[\cdot, \cdot]_0$ always denotes the one in (4.8), i.e., it corresponds to *precise data* y; accordingly, the polynomials $\{p_k^{[n]}\}$ are always residual polynomials of conjugate gradient type methods with right-hand side y. Note that this notation differs from the other sections where these quantities always referred to the actual (i.e., perturbed) right-hand side. The reason for this change of notation is the fact that all residual polynomials that are of interest below can easily be expressed in terms of $p_k^{[n]}$.

For $\delta > 0$ consider the following perturbation y^δ of y:

$$y^\delta(t) = \begin{cases} \delta, & t = 0, \\ y(t), & 0 < t < 1. \end{cases} \tag{4.9}$$

Obviously, y^δ does not belong to $\mathcal{R}(T)$ and, cf. (4.4),

$$\|y - y^\delta\| = \delta.$$

The residual polynomials of CG with right-hand side y^δ will be denoted by $\{p_k^\delta\}$. They are orthogonal with respect to the inner product $[\cdot, \cdot]^\delta$ given by

$$[\varphi, \psi]^\delta = \delta^2\varphi(0)\psi(0) + [\varphi, \psi]_0. \tag{4.10}$$

Proposition 4.5 *The residual polynomials p_k^δ of CG corresponding to y^δ of (4.9) equal*

$$p_k^\delta = p_k^{[0]} - \vartheta_k \lambda p_{k-1}^{[2]}, \qquad k \geq 1,$$

where

$$\vartheta_k = \delta^2 [p_{k-1}^{[2]}, p_{k-1}^{[2]}]_1^{-1}. \tag{4.11}$$

Proof. The statement is proved by checking the orthogonality relations: for $0 < j < k$ one has

$$[p_k^\delta, \lambda^j]^\delta = [p_k^\delta, \lambda^j]_0 = [p_k^{[0]}, \lambda^j]_0 - \vartheta_k[p_{k-1}^{[2]}, \lambda^{j-1}]_2 = 0 \, ;$$

on the other hand, for $k > 0$ and $j = 0$ one has

$$[p_k^\delta, \lambda^j]^\delta = [p_k^\delta, 1]^\delta = \delta^2 p_k^\delta(0) + [p_k^{[0]}, 1]_0 - \vartheta_k[p_{k-1}^{[2]}, 1]_1 \, ,$$

and the right-hand side vanishes if $\vartheta_k = \delta^2[p_{k-1}^{[2]}, 1]_1^{-1}$. Therefore, (4.11) follows from (2.23). □

The following lemma states a few asymptotic properties of the residual polynomials under consideration; a more refined statement of some of these asymptotics is given in Lemma 5.3 in Section 5.2.

Lemma 4.6 *Let $[\cdot, \cdot]_0$ be as in (4.8). For fixed $n \in \mathbf{N}_0$ the following asymptotics hold as $k \to \infty$:*

(i) $\quad [p_k^{[n]}, p_k^{[n]}]_n \sim k^{-4\nu-3-2n}$,

(ii) $\quad [p_k^{[n+1]}, p_k^{[n+1]}]_n \sim k^{-4\nu-4-2n}$,

(iii) $\quad |p_k^{[n]\prime}(0)| \sim k^2$,

(iv) $\quad \pi_{k,n} \sim k$.

Here, $\pi_{k,n}$ is defined as in Proposition 2.5.

Proof. The assertions follow readily from well-known results concerning Jacobi polynomials, cf. [76, Chapter 4]. Denote by $u_k^{[n]}$ an orthonormal multiple of $p_k^{[n]}$. It is obvious that

$$[p_k^{[n]}, p_k^{[n]}]_n = |u_k^{[n]}(0)|^{-2}[u_k^{[n]}, u_k^{[n]}]_n = |u_k^{[n]}(0)|^{-2} \, . \tag{4.12}$$

Therefore, the growth rate (i) follows from

$$|u_k^{[n]}(0)| \sim k^{2\nu+3/2+n} \, , \tag{4.13}$$

cf. Equations (4.1.1) and (4.3.4) in [76].
    According to (2.21),

$$[p_k^{[n+1]}, p_k^{[n+1]}]_n^{-1} = \sum_{j=0}^{k} [p_j^{[n]}, p_j^{[n]}]_n^{-1} \, ,$$

and hence, (ii) follows from (i).
    Finally, the asymptotics (iii) and (iv) of $p_k^{[n]\prime}(0)$ and $\pi_{k,n}$, respectively, follow from a well-known identity for the derivative of Jacobi polynomials, cf. [76, (4.21.7)]. □

62

In the sequel it will be shown that the norm of the residual

$$r_k^\delta = y^\delta - T x_k^\delta \tag{4.14}$$

will never reach the tolerance $\tau\delta$, provided δ is sufficiently small. In fact, by Proposition 4.5,

$$
\begin{aligned}
\|r_k^\delta\|^2 &= [p_k^\delta, p_k^\delta]^\delta \\
&= \delta^2 + [p_k^{[0]} - \vartheta_k \lambda p_{k-1}^{[2]}, p_k^{[0]} - \vartheta_k \lambda p_{k-1}^{[2]}]_0 \\
&= \delta^2 + [p_k^{[0]}, p_k^{[0]}]_0 - 2\vartheta_k [p_k^{[0]}, \lambda p_{k-1}^{[2]}]_0 + \vartheta_k^2 [p_{k-1}^{[2]}, p_{k-1}^{[2]}]_2 .
\end{aligned}
$$

Using Propositions 2.5 and 2.8, together with orthogonality relations, one obtains

$$
\begin{aligned}
[p_k^{[0]}, \lambda p_{k-1}^{[2]}]_0 &= \frac{1}{\pi_{k-1,1}} [p_k^{[0]}, p_{k-1}^{[1]} - p_k^{[1]}]_0 \\
&= -\frac{1}{\pi_{k-1,1}} ([p_k^{[0]} - p_k^{[1]}, p_k^{[1]}]_0 + [p_k^{[1]}, p_k^{[1]}]_0) \\
&= -\frac{1}{\pi_{k-1,1}} (-\theta_{k,0} [p_{k-1}^{[2]}, p_k^{[1]}]_1 + [p_k^{[1]}, p_k^{[1]}]_0) \\
&= -\frac{1}{\pi_{k-1,1}} [p_k^{[1]}, p_k^{[1]}]_0 ,
\end{aligned}
$$

and hence,

$$
\|r_k^\delta\|^2 = \delta^2 + [p_k^{[0]}, p_k^{[0]}]_0 + \frac{2\vartheta_k}{\pi_{k-1,1}} [p_k^{[1]}, p_k^{[1]}]_0 + \vartheta_k^2 [p_{k-1}^{[2]}, p_{k-1}^{[2]}]_2 .
$$

Inserting (4.11) and the asymptotics in Lemma 4.6, it follows that

$$
\|r_k^\delta\|^2 \sim \delta^2 + k^{-4\nu-3}(1 + \delta^2 k^{4\nu+4} + \delta^4 k^{8\nu+8}) .
$$

The right-hand side becomes minimal for $k \sim \delta^{-1/2\nu+2}$, hence

$$\min_{k \in \mathbb{N}_0} \|r_k^\delta\|^2 \sim \delta^2 (1 + \delta^{-1/2\nu+2}) . \tag{4.15}$$

It follows that Stopping Rule 3.10, i.e., the discrepancy principle, does not halt the iteration when δ is sufficiently small. Nevertheless, CG diverges according to Theorem 4.4.

The above argument is easily modified to provide a counterexample for CGME: one simply takes T to be the square root of the operator in (4.5) without changing y (of course, the solution x is then different). In this case, the inner products and the residual polynomials are the same as above, leading thus to the same residuals.

One might argue that the failure of the discrepancy principle is due to the fact that $y - y^\delta$ is orthogonal to $\mathcal{R}(T)$. This is not the reason, though, as the following modification shows, where both y and y^δ belong to $\mathcal{R}(T)$. Let $K : \mathcal{X}_K \to \mathcal{X}_K$ be a compact operator in some Hilbert space $\mathcal{X}_K$ (let K be selfadjoint, positive definite and nondegenerate). Consider the product space $\mathcal{X}_K \oplus \mathcal{X}$ with the canonical embeddings; no notational care will be taken to distinguish elements in one of the factor spaces from the corresponding element in the product space. With T as before, define

$$T_K : \mathcal{X}_K \oplus \mathcal{X} \longrightarrow \mathcal{X}_K \oplus \mathcal{X}, \qquad T_K : \begin{cases} x \mapsto Kx, & x \in \mathcal{X}_K, \\ x \mapsto Tx, & x \in \mathcal{X}. \end{cases}$$

Obviously, the spectrum of T_K coincides with the spectrum of T up to an additional sequence of countably many eigenvalues $\lambda_j > 0$ with eigenvectors $v_j \in \mathcal{X}_K$, $\|v_j\| = 1$. Let $y = Tx^{(\nu)}$ as before, and for $\delta > 0$ and $j \in \mathbb{N}$ define

$$y^{\delta,j}(t) = \delta v_j \oplus \begin{cases} 0, & 0 \le t < \lambda_j, \\ y(\frac{t - \lambda_j}{1 - \lambda_j}), & \lambda_j \le t \le 1 \end{cases} \quad \in \mathcal{R}(T_K).$$

Obviously,
$$\|y - y^{\delta,j}\| = \delta + o(1), \qquad j \to \infty. \tag{4.16}$$

Denote the spectral family of T_K by $\{F_\lambda\}$. While $d\|F_\lambda y\|^2$ defines the same inner product as in the original example, cf. (4.8), the spectral distribution $d\|F_\lambda y^{\delta,j}\|^2$ defines the "translated" inner product

$$\begin{aligned}
[\varphi, \psi]^{\delta,j} &:= \int_0^1 \varphi(\lambda)\psi(\lambda)\, d\|F_\lambda y^{\delta,j}\|^2 \\
&= [\varphi(\lambda_j + (1 - \lambda_j)\cdot), \psi(\lambda_j + (1 - \lambda_j)\cdot)]^\delta;
\end{aligned} \tag{4.17}$$

here, $[\cdot, \cdot]^\delta$ is as in (4.10). It follows that the residual polynomials $\{p_k^{\delta,j}\}$ of CG for $T_K x = y^{\delta,j}$ are given by

$$p_k^{\delta,j}(\lambda) = p_k^\delta(\frac{\lambda - \lambda_j}{1 - \lambda_j}) / p_k^\delta(\frac{-\lambda_j}{1 - \lambda_j}) \quad \in \Pi_k^0. \tag{4.18}$$

Let $\{x_k^{\delta,j}\}$ denote the CG iterates; looking only at their component in $\mathcal{X}_K$, one obtains the following lower bound for the iteration error:

$$\|x^{(\nu)} - x_k^{\delta,j}\| \ge (1 - p_k^{\delta,j}(\lambda_j)) \frac{\delta}{\lambda_j} ;$$

writing ε for $\lambda_j/(1 - \lambda_j)$, using (4.18) and the mean-value inequality, this yields

$$\|x^{(\nu)} - x_k^{\delta,j}\| \ge \left(1 - \frac{1}{p_k^\delta(-\varepsilon)}\right) \frac{\delta}{\lambda_j} = \left(1 - \frac{1}{1 + \varepsilon |p_k^{\delta\prime}(-\tilde{\varepsilon})|}\right) \frac{\delta}{\lambda_j} = \frac{\varepsilon |p_k^{\delta\prime}(-\tilde{\varepsilon})|}{1 + \varepsilon |p_k^{\delta\prime}(-\tilde{\varepsilon})|} \frac{\delta}{\lambda_j}$$

for some $\tilde{\varepsilon}$ with $0 < \tilde{\varepsilon} < \varepsilon$. Using the convexity of p_k^δ in $\mathbb{R}^-$, it follows that

64

$$\|x^{(\nu)} - x_k^{\delta,j}\| \geq \frac{|p_k^{\delta\prime}(0)|}{1 + \varepsilon|p_k^{\delta\prime}(0)|} \frac{\varepsilon\delta}{\lambda_j} = \frac{|p_k^{\delta\prime}(0)|}{1 + \varepsilon|p_k^{\delta\prime}(0)|} \frac{\delta}{1 - \lambda_j}. \tag{4.19}$$

Furthermore, by definition of p_k^δ, cf. Proposition 4.5, and by Lemma 4.6, there exists $c > 0$ such that

$$|p_k^{\delta\prime}(0)| = |p_k^{[0]\prime}(0) - \vartheta_k| \geq |p_k^{[0]\prime}(0)| \geq 2ck^2, \qquad k \in \mathbf{N}.$$

Given $k_0 \in \mathbf{N}$, the monotonicity of $p_k^{\delta\prime}(0)$ (as a sequence in k) therefore yields

$$\frac{|p_k^{\delta\prime}(0)|}{1 + \varepsilon|p_k^{\delta\prime}(0)|} \frac{\delta}{1 - \lambda_j} \geq \frac{|p_{k_0}^{\delta\prime}(0)|}{1 + \varepsilon|p_{k_0}^{\delta\prime}(0)|} \frac{\delta}{1 - \lambda_j} \geq ck_0^2\delta, \qquad k \geq k_0,$$

provided $j = j(k_0)$ is sufficiently large. Inserting the above into (4.19) finally leads to

$$\|x^{(\nu)} - x_k^{\delta,j}\| \geq ck_0^2\delta \qquad \text{for } k \geq k_0 \text{ and } j \geq j(k_0). \tag{4.20}$$

Recall the definition of r_k^δ as given in (4.14), and denote by $\{r_k^{\delta,j}\}$ the residuals of the CG iterates with respect to the modified problem $T_K x = y^{\delta,j}$; from (4.17) and (4.18) follows

$$\|r_k^{\delta,j}\|^2 = [p_k^{\delta,j}, p_k^{\delta,j}]^{\delta,j} = |p_k^\delta(\frac{-\lambda_j}{1 - \lambda_j})|^{-2}[p_k^\delta, p_k^\delta]^\delta = |p_k^\delta(\frac{-\lambda_j}{1 - \lambda_j})|^{-2}\|r_k^\delta\|^2,$$

and hence, for fixed k,

$$\|r_k^{\delta,j}\|^2 \to \|r_k^\delta\|^2, \qquad j \to \infty. \tag{4.21}$$

Consider now Stopping Rule 3.10 for some fixed $\tau > 1$. As shown in (4.15), $\|r_k^\delta\|$ never drops below the tolerance $3\tau\delta$, if $\delta \leq \delta_0$ with some δ_0 sufficiently small. Fix $\delta \leq \delta_0$ and let $k_0 = k_0(\delta)$ be given. In view of (4.16) and (4.21) one can choose $j_0 = j_0(\delta) \geq j(k_0)$ (with $j(k_0)$ as in (4.20)) such that $\|y - y^{\delta,j_0}\| \leq 2\delta$ and

$$\|r_k^{\delta,j_0}\| > \frac{2}{3}\|r_k^\delta\| \geq 2\tau\delta, \qquad 0 \leq k \leq k_0.$$

The conclusion is the following: since $y^{\delta,j} \in \mathcal{R}(T)$, the residuals $r_k^{\delta,j}$ converge to zero as $k \to \infty$ by Theorem 4.2; therefore, the discrepancy principle terminates the iteration for some finite stopping index $k(\delta)$. Obviously, $k(\delta) \geq k_0$, hence (4.20) yields

$$\|x^{(\nu)} - x_{k(\delta)}^{\delta,j_0}\| \geq ck_0^2\delta.$$

Choosing $k_0 = k_0(\delta) \sim \delta^{-1}$, this shows that $\|x_{k(\delta)}^{\delta,j_0(\delta)}\| \to \infty$ as $\delta \to 0$. In other words, although the discrepancy principle terminates the iteration, the approximations diverge as $\delta \to 0$.

4.3 An order-optimal stopping rule

The failure of the discrepancy principle for CG and CGME may be understood as a problem of saturation. It is well-known from other regularization methods that a successful parameter choice on the basis of the discrepancy principle is only possible if the residuals converge with an appropriate order of convergence. For example, in Tikhonov regularization this is the case if $x \in \mathcal{R}(|T|)$ but not if $x \in \mathcal{R}(|T|^2)$. The reason lies in a certain *saturation* of the rate with which the residuals converge to zero.

To construct an order-optimal stopping rule one has to monitor a sequence $\{\|\varphi_k(T)y^\delta\|\}$ with functions φ_k that do not share this saturation property. Maybe the most straightforward choice is $\varphi_k = p_k^{[1]}$, in which case $\|\varphi_k(T)y^\delta\| = [p_k^{[1]}, p_k^{[1]}]_0^{1/2}$, i.e., the norm of the kth residual of MR (CGNE).

Stopping Rule 4.7 *Fix $\tau > 1$ and assume $\|y - y^\delta\| \leq \delta$. Terminate the CG (CGME) iteration as soon as $\|y^\delta - Tx_k^\delta\| = 0$, or when for the first time*

$$\sum_{j=0}^{k} \|y^\delta - Tx_j^\delta\|^{-2} \geq (\tau\delta)^{-2}.$$

Denote the resulting stopping index by $k(\delta)$.

Remark. Note that it follows from Lemma 2.4 that $k(\delta)$ is the smallest integer k, for which $[p_k^{[1]}, p_k^{[1]}]_0^{1/2} \leq \tau\delta$. In other words, the stopping index of Stopping Rule 4.7 coincides with the stopping index for MR (CGNE) as determined by the discrepancy principle. Consequently, one always has $k(\delta) \leq \kappa$, and hence the iteration is terminated before a breakdown can occur.

It has to be emphasized that the norm of the residual $y^\delta - Tx_k^\delta$ is computed anyway in CG and CGME, cf. Algorithms 2.2 and 2.4. In the following it will be shown that this stopping rule makes CG and CGME order-optimal regularization methods.

Theorem 4.8 *Let y satisfy Assumption 3.6 and let $\|y - y^\delta\| \leq \delta$. If CG or CGME is applied to y^δ and terminated after $k(\delta)$ iterations according to Stopping Rule 4.7, then there exists some uniform constant $c > 0$ such that*

$$\|T^\dagger y - x_{k(\delta)}^\delta\| \leq c\omega^{1/\mu+1}\delta^{\mu/\mu+1}.$$

Proof. In the sequel, k always denotes the stopping index $k(\delta)$ as determined by Stopping Rule 4.7. CG will be considered first. Denoting by z_k^δ the corresponding MR approximation after k steps, one has

$$\|T^\dagger y - x_k^\delta\| \leq \|T^\dagger y - z_k^\delta\| + \|z_k^\delta - x_k^\delta\| . \tag{4.22}$$

By the remark following Stopping Rule 4.7, the stopping index k is the same as the one chosen by the discrepancy principle for MR. Therefore, the first term on the right-hand side of (4.22) satisfies the asserted bound by virtue of Theorem 3.11, and it remains to estimate $x_k^\delta - z_k^\delta$.

Assume $x_k^\delta \neq z_k^\delta$. This implies that $k \neq 0$, and that the residual polynomials $p_k^{[0]}$ and $p_k^{[1]}$ differ. In view of the remark following Proposition 2.8, (2.24) applies to the present situation: denoting by $\{q_k^{[0]}\}$ and $\{q_k^{[1]}\}$ the iteration polynomials of CG and MR, respectively, $x_k^\delta - z_k^\delta$ can be rewritten as

$$x_k^\delta - z_k^\delta = q_{k-1}^{[0]}(T)y^\delta - q_{k-1}^{[1]}(T)y^\delta = \theta_{k,0}\, p_{k-1}^{[2]}(T)y^\delta .$$

Thus, it follows from Corollary 2.9 and Proposition 2.1 that

$$\|x_k^\delta - z_k^\delta\|^2 = \frac{[p_k^{[1]}, p_k^{[1]}]_0^2}{[p_{k-1}^{[2]}, p_{k-1}^{[2]}]_1^2}\, [p_{k-1}^{[2]}, p_{k-1}^{[2]}]_0 \leq \left(\frac{[p_{k-1}^{[2]}, p_{k-1}^{[2]}]_0}{[p_{k-1}^{[2]}, p_{k-1}^{[2]}]_1}\right)^2 [p_k^{[1]}, p_k^{[1]}]_0 . \tag{4.23}$$

Let $\varepsilon = (c_\varepsilon \delta/\omega)^{1/\mu+1}$, with c_ε to be chosen later. As has been shown in (3.23) in the proof of Theorem 3.11, one has

$$0 \leq p_{k-1}^{[2]}(\lambda) \leq 1, \qquad 0 \leq \lambda \leq \varepsilon ,$$

provided c_ε is sufficiently small. Therefore one can proceed as in Step 2 of the proof of Theorem 3.11 to obtain

$$\begin{aligned}
[p_{k-1}^{[2]}, p_{k-1}^{[2]}]_0 &= \|E_\varepsilon p_{k-1}^{[2]}(T)y^\delta\|^2 + \|(I - E_\varepsilon)p_{k-1}^{[2]}(T)y^\delta\|^2 \\
&\leq (\delta + \varepsilon^{\mu+1}\omega)^2 + \varepsilon^{-1}\|p_{k-1}^{[2]}(T)T^{1/2}y^\delta\|^2 \\
&= (1 + c_\varepsilon)^2 \delta^2 + \varepsilon^{-1}[p_{k-1}^{[2]}, p_{k-1}^{[2]}]_1 .
\end{aligned}$$

Because of the optimality property of $p_{k-1}^{[1]}$ and the equivalence of Stopping Rule 4.7 with the discrepancy principle for MR,

$$\delta^2 \leq \frac{1}{\tau^2}\, [p_{k-1}^{[1]}, p_{k-1}^{[1]}]_0 \leq \frac{1}{\tau^2}\, [p_{k-1}^{[2]}, p_{k-1}^{[2]}]_0 .$$

Inserting this into the previous estimate one obtains

$$(1 - \frac{(1 + c_\varepsilon)^2}{\tau^2})[p_{k-1}^{[2]}, p_{k-1}^{[2]}]_0 \leq \varepsilon^{-1}[p_{k-1}^{[2]}, p_{k-1}^{[2]}]_1 .$$

Therefore, by choosing $0 < c_\varepsilon < \tau - 1$ sufficiently small, one can find some $c > 0$ such that

67

$$\frac{[p_{k-1}^{[2]}, p_{k-1}^{[2]}]_0}{[p_{k-1}^{[2]}, p_{k-1}^{[2]}]_1} \leq c(\omega/\delta)^{1/\mu+1} . \tag{4.24}$$

Insertion of (4.24) into (4.23) provides

$$\|x_k^\delta - z_k^\delta\| \leq c(\omega/\delta)^{1/\mu+1}[p_k^{[1]}, p_k^{[1]}]_0^{1/2} ,$$

and, since $[p_k^{[1]}, p_k^{[1]}]_0^{1/2} \leq \tau\delta$ by the definition of $k = k(\delta)$, this yields

$$\|x_k^\delta - z_k^\delta\| \leq c\tau\omega^{1/\mu+1}\delta^{\mu/\mu+1} .$$

Thus, the statement of the theorem follows from (4.22).

Consider CGME next, and denote this time by $\{z_k^\delta\}$ the iterates of CGNE. The stopping index $k = k(\delta)$ now agrees with the one obtained by the discrepancy principle for CGNE and, as above, cf. (4.22), it remains to estimate $x_k^\delta - z_k^\delta$. This time one has

$$x_k^\delta - z_k^\delta = T^* q_{k-1}^{[0]}(TT^*)y^\delta - T^* q_{k-1}^{[1]}(TT^*)y^\delta = \theta_{k,0} T^* p_{k-1}^{[2]}(TT^*)y^\delta ,$$

and therefore Corollary 2.9 and Proposition 2.1 yield

$$\|x_k^\delta - z_k^\delta\|^2 = \frac{[p_k^{[1]}, p_k^{[1]}]_0^2}{[p_{k-1}^{[2]}, p_{k-1}^{[2]}]_1^2} [p_{k-1}^{[2]}, p_{k-1}^{[2]}]_1 \leq \frac{[p_{k-1}^{[2]}, p_{k-1}^{[2]}]_0}{[p_{k-1}^{[2]}, p_{k-1}^{[2]}]_1} [p_k^{[1]}, p_k^{[1]}]_0 \tag{4.25}$$

in the CGME case. This is the analog of (4.23) and the remaining part of the proof is now clear, cf. [35] for more details. □

In particular, this result shows that CG and CGME are regularization methods, when combined with the above stopping rule:

Theorem 4.9 *Let $y \in \mathcal{R}(T)$ and $\|y - y^\delta\| \leq \delta$. If the stopping index $k(\delta)$ is determined according to Stopping Rule 4.7 then $x_{k(\delta)}^\delta \to T^\dagger y$ as $\delta \to 0$.*

The proof is the same, word by word, as the proof of Theorem 3.12. In fact, this argument is valid for any order-optimal regularization method (cf. [67]).

4.4 A heuristic stopping rule

As in Section 3.4, the sharp estimates for the iteration error can also be used to design heuristic stopping rules for CG and CGME. In contrast to the order-optimal rule from the previous section, these rules define different stopping indices than the corresponding rules of Section 3.4 for MR and CGNE, respectively.

Stopping Rule 4.10 *Let $\eta_0 = \|y^\delta\|$, and for $k \geq 1$ compute*

$$\eta_k = \begin{cases} 0 & \|y^\delta - Tx_k^\delta\| = 0, \\ |p_k'(0)| \, (\sum_{j=0}^k \|y^\delta - Tx_j^\delta\|^{-2})^{-1/2}, & \|y^\delta - Tx_k^\delta\| \neq 0, \end{cases} \quad (4.26)$$

in the case of CG, *and*

$$\eta_k = \begin{cases} 0 & \|y^\delta - Tx_k^\delta\| = 0, \\ |p_k'(0)|^{1/2}(\sum_{j=0}^k \|y^\delta - Tx_j^\delta\|^{-2})^{-1/2}, & \|y^\delta - Tx_k^\delta\| \neq 0, \end{cases}$$

for CGME, *respectively. In either case terminate the iteration after $k(y^\delta)$ steps provided $\eta_{k(y^\delta)} \leq \eta_k$ for all $0 \leq k \leq \kappa$.*

Like the heuristic rules in Section 3.4 Stopping Rule 4.10 may fail since η_k can tend to zero as $k \to \infty$. The following result provides an estimate for the approximation in case of a finite stopping index.

Theorem 4.11 *Let y satisfy Assumption 3.6 and let $\|y - y^\delta\| \leq \|y\|$. If Stopping Rule 4.10 determines a finite stopping index $k(y^\delta)$, and if $\eta_{k(y^\delta)} \neq 0$ then the following a posteriori error bound holds:*

$$\|T^\dagger y - x_{k(y^\delta)}^\delta\| \leq c(1 + \frac{\|y - y^\delta\|}{\delta_*})\omega^{1/\mu+1}\rho_*^{\mu/\mu+1},$$

where

$$\delta_* = \Big(\sum_{j=0}^{k(y^\delta)} \|y^\delta - Tx_j^\delta\|^{-2} \Big)^{-1/2} \quad and \quad \rho_* = \max\{\|y - y^\delta\|, \delta_*\}.$$

Proof. The proof is only given for CG, and without loss of generality it will be assumed that $\|y - y^\delta\| = \delta.$ As in the proof of Theorem 4.8 $\|x_k^\delta - z_k^\delta\|$ can be estimated by (4.23) provided $k \geq 1$. Thus, Lemma 3.8 and (4.22) imply that

$$\|T^\dagger y - x_k^\delta\| \leq c(\omega^{1/\mu+1}\rho_k^{\mu/\mu+1} + |p_k^{[1]\prime}(0)| \, \delta + \frac{[p_{k-1}^{[2]}, p_{k-1}^{[2]}]_0}{[p_{k-1}^{[2]}, p_{k-1}^{[2]}]_1} \rho_k) \quad (4.27)$$

with ρ_k defined as in (3.12), namely $\rho_k = \max\{[p_k^{[1]}, p_k^{[1]}]_0^{1/2}, \delta\}$. It follows from (2.21) that $\rho_{k(y^\delta)} = \rho_*$. In the following, the fraction on the right-hand side of (4.27) will be investigated. By orthogonality,

$$[p_{k-1}^{[2]}, p_{k-1}^{[2]}]_0 = [p_{k-1}^{[2]}, p_{k-1}^{[2]} - p_k^{[0]}]_0 = [p_{k-1}^{[2]}, (p_{k-1}^{[2]} - p_k^{[0]})'(0)\lambda + \lambda^2 s_{k-2}]_0,$$

where s_{k-2} is a polynomial of degree $k - 2$. This yields, cf. (2.23),

$$[p^{[2]}_{k-1}, p^{[2]}_{k-1}]_0 = (p^{[2]}_{k-1} - p^{[0]}_k)'(0) \, [p^{[2]}_{k-1}, 1]_1 + [p^{[2]}_{k-1}, s_{k-2}]_2 = (p^{[2]}_{k-1} - p^{[0]}_k)'(0) \, [p^{[2]}_{k-1}, p^{[2]}_{k-1}]_1 .$$

In other words,

$$\frac{[p^{[2]}_{k-1}, p^{[2]}_{k-1}]_0}{[p^{[2]}_{k-1}, p^{[2]}_{k-1}]_1} = p^{[2]}_{k-1}{}'(0) - p^{[0]}_k{}'(0) . \tag{4.28}$$

From (2.19) and Corollary 2.7 follows that $|p^{[2]}_{k-1}{}'(0)| < |p^{[1]}_k{}'(0)| \le |p^{[0]}_k{}'(0)|$; hence, inserting (4.28) into (4.27) one obtains for $k = k(y^\delta)$ (provided, for the moment, that $k(y^\delta) \ge 1$):

$$\|T^\dagger y - x^\delta_{k(y^\delta)}\| \le c(\omega^{1/\mu+1} \rho_*^{\mu/\mu+1} + |p^{[0]}_{k(y^\delta)}{}'(0)| \rho_*) . \tag{4.29}$$

If $k(y^\delta) = 0$ then (4.29) follows immediately from the interpolation inequality (3.5) which states that

$$\|T^\dagger y - x^\delta_0\| = \|T^\dagger y\| \le \omega^{1/\mu+1} \rho_0^{\mu/\mu+1} ,$$

and ρ_0 coincides with ρ_* in this case. It follows that (4.29) holds true whatever stopping index $k(y^\delta)$ is chosen by Stopping Rule 4.10.

The remainder of the proof is similar to the proof of Theorem 3.14. From (4.29) and the definition (4.26) of $\eta_{k(y^\delta)}$ follows

$$\|T^\dagger y - x^\delta_{k(y^\delta)}\| \le c(\omega^{1/\mu+1} \rho_*^{\mu/\mu+1} + \frac{\rho_*}{\delta_*} \eta_{k(y^\delta)}) \le c(\omega^{1/\mu+1} \rho_*^{\mu/\mu+1} + \frac{\rho_*}{\delta_*} \eta_{k(\delta)}) , \tag{4.30}$$

where $k(\delta)$ is the stopping index corresponding to Stopping Rule 4.7. For $k > 0$ Propositions 2.5, 2.8, and (4.28) imply

$$|p^{[0]}_k{}'(0)| = p^{[2]}_{k-1}{}'(0) - p^{[0]}_k{}'(0) - \theta_{k,1} - \pi_{k-1,2} + |p^{[1]}_k{}'(0)| \le |p^{[1]}_k{}'(0)| + \frac{[p^{[2]}_{k-1}, p^{[2]}_{k-1}]_0}{[p^{[2]}_{k-1}, p^{[2]}_{k-1}]_1} .$$

Hence, using (3.18) and (4.24) (recall that $k = k(\delta)$ is also the stopping index obtained by the discrepancy principle for MR), one obtains

$$|p^{[0]}_{k(\delta)}{}'(0)| \le c \, (\omega/\delta)^{1/\mu+1} , \tag{4.31}$$

which yields

$$\eta_{k(\delta)} \le c\tau \, \omega^{1/\mu+1} \delta^{\mu/\mu+1} . \tag{4.32}$$

For $k(\delta) = 0$, on the other hand, (4.32) follows as in the proof of Theorem 3.14, cf. (3.28). Inserting (4.32) into (4.30) the theorem follows since $\rho_*/\delta_* = \max\{1, \delta/\delta_*\}$.

$\square$

Remark. For CGME, two of the factors on the right-hand side of (4.27) come with a square root: this follows from the remark following Corollary 3.9, and from (4.25) which replaces (4.23) for CGME. The remaining modifications are straightforward.

Corollary 3.15 and Theorem 3.16 carry over to the present situation without any significant change of proof, except that (4.31) instead of (3.18) has to be used in the proof of Theorem 3.16. The details are left to the reader.

Notes and remarks

Section 4.1. KING [47] has shown that CGME converges for precise data. LARDY [51] comments on the behaviour of the iterates when $y \notin \mathcal{R}(T)$.

Section 4.2. In view of the fact that CG (CGME) achieves order-optimal accuracy (as established in Section 4.3), the failure of the discrepancy principle is surprising; following a general methodology by RAUS [69] – although developed for a different class of regularization methods – one would expect the discrepancy principle to be an adequate stopping rule. The particular counterexample constructed in Section 4.2 is taken from [35]. The orthogonal polynomials used for this counterexample were analyzed by KOORNWINDER [48]. Of course, Proposition 4.5 holds for any inner product $[\cdot, \cdot]_0$, compare CHIHARA [10].

Section 4.3. The material in this section is taken from [35]. The saturation of Tikhonov regularization and the resulting suboptimality of the discrepancy principle has been studied by GROETSCH [30, Theorem 3.3.6]. The same phenomenon has been observed for the ν-methods, cf. [37]; in [37] this has been remedied by using the residual polynomials corresponding to the $\nu + 1/2$-method for the construction of an order-optimal stopping rule. This is a similar idea as has been used here.

Section 4.4. In principle one can alternatively extend the heuristic stopping rules considered by HANSEN [42] for CGNE to CG (CGME). It does not seem to be clear, though, whether the saturated decay rate of the residuals of CG (CGME) will affect the behaviour of these stopping rules.

5. On the Number of Iterations

The efficiency of iterative regularization methods – besides their accuracy – depends on the number of required iterations to meet the stopping criterion. In this chapter, in particular in the first section, several estimates are given for the stopping index which depend on the spectral distribution of the unperturbed right-hand side. It becomes clear that conjugate gradient type methods are especially efficient when T is a compact operator. Section 5.2 presents a refined comparison of MR and CG in a case study, namely for the example constructed in Section 4.2. Recall that Stopping Rules 3.10 and 4.7, respectively, terminate the two schemes after precisely the same number of iterations, and the accuracy of the two approximations is similar, at least in terms of powers of δ. The final section presents some numerical results for an ill-posed image reconstruction problem.

5.1 General estimates for the stopping index

Consider the stopping indices $k(\delta)$ as determined by Stopping Rule 3.10 (Stopping Rule 4.7) for conjugate gradient type methods with parameter $n \geq 1$ ($n = 0$) including, in particular, MR, respectively CG. In the following, three bounds for the asymptotic growth of $k(\delta)$ as $\delta \to 0$ will be derived. As will be seen, these bounds are independent of the parameter n; however, it is clear, that MR will require fewest iterations to meet the stopping criterion, cf. Proposition 2.1.

Theorem 5.1 *Let $\|y - y^\delta\| \leq \delta$ and $k(\delta)$ be the stopping index for any of the conjugate gradient type methods of Section 2.1 as determined by Stopping Rule 3.10 and 4.7 for $n \geq 1$ and $n = 0$, respectively. If y satisfies Assumption 3.6 then*

$$k(\delta) \leq c \, (\omega/\delta)^{1/2\mu+2} \,, \tag{5.1}$$

and this estimate is sharp in the sense that the exponent cannot be replaced by a smaller one in general.

Proof. Let n be the parameter of the conjugate gradient type method under consideration. Recall that the stopping indices coincide for $n = 0$ and $n = 1$, see the remark following Stopping Rule 4.7. Consider this case first, i.e., let $n = 1$. By definition, the iteration is terminated when, for the first time,

$$[p_k^{[1]}, p_k^{[1]}]_0 \leq \tau^2 \delta^2$$

for some fixed $\tau > 1$. Let $k := k(\delta) - 1$. For $\nu > 0$ consider the Jacobi polynomial

$$\varphi_k(\lambda) := P_k^{(\nu-1/2,-1/2)}(1 - 2\lambda)/P_k^{(\nu-1/2,-1/2)}(1).\tag{5.2}$$

According to [76, Sect. 7.32], φ_k satisfies the following inequalities:

$$|\varphi_k(\lambda)| \leq 1, \qquad |\lambda^\nu \varphi_k^2(\lambda)| \leq c(k+1)^{-2\nu},\tag{5.3}$$

uniformly for $0 \leq \lambda \leq 1$ and $k \in \mathbf{N}_0$. Using $\nu = 2\mu + 2$ this yields

$$\|\varphi_k(T)y^\delta\| \leq \|\varphi_k(T)(y - y^\delta)\| + \|\varphi_k(T)T^{\mu+1}w\| \leq \delta + c(k+1)^{-2\mu-2}\omega.$$

Since $k = k(\delta) - 1$ Proposition 2.1 implies

$$\tau\delta \leq [p_k^{[1]}, p_k^{[1]}]_0^{1/2} \leq [\varphi_k, \varphi_k]_0^{1/2} \leq \delta + ck(\delta)^{-2\mu-2}\omega,$$

from which (5.1) follows for $n = 1$ and $n = 0$.

Consider now the case $n \geq 2$. By virtue of (3.20) there exists $c > 0$ such that

$$[p_k^{[n]}, p_k^{[n]}]_{n-1} \geq c\delta^2 (\delta/\omega)^{(n-1)/\mu+1}.\tag{5.4}$$

If φ_k is chosen as in (5.2) with $\nu = n + 2\mu + 1$ then (5.3) implies that

$$|\lambda^\alpha \varphi_k^2(\lambda)| \leq c(k+1)^{-2\alpha}, \qquad 0 \leq \lambda \leq 1,\ k \in \mathbf{N}_0,$$

for every $0 \leq \alpha \leq \nu$, in particular for $\alpha = n - 1$. To obtain these bounds for $\alpha < \nu$, one applies the first inequality in (5.3) for $\lambda \in [0, (k+1)^{-2}]$, and the second inequality in (5.3) for the remaining interval $[(k+1)^{-2}, 1]$. It follows that

$$\begin{aligned}\|T^{(n-1)/2}\varphi_k(T)y^\delta\| &\leq \|T^{(n-1)/2}\varphi_k(T)(y^\delta - y)\| + \|\varphi_k(T)T^{(n+1+2\mu)/2}w\| \\ &\leq c((k+1)^{-n+1}\delta + (k+1)^{-n-2\mu-1}\omega).\end{aligned}$$

Using (5.4) and the optimality property stated in Proposition 2.1, this yields

$$(\delta/\omega)^{(n-1)/2\mu+2} \leq c\,k(\delta)^{-n+1}(1 + k(\delta)^{-2\mu-2}\omega/\delta).$$

Let $l = k(\delta)^{2\mu+2}\delta/\omega$; then the above inequality can be rewritten as

$$l^{(n-1)/2\mu+2} \leq c(1 + l^{-1}),$$

and since $n > 1$ this implies a finite upper bound for l; in other words, (5.1) holds for $n \geq 2$ as well.

To establish the sharpness of (5.1) consider the example (4.7) of Section 4.2 with no perturbation, i.e., $y^\delta = y = Tx^{(\nu)}$ with $x^{(\nu)}$ as in (4.6). By virtue of Lemma 4.6 (recall that $n \geq 1$),

$$[p_k^{[n]}, p_k^{[n]}]_{n-1} \sim k^{-4\nu-2-2n}, \qquad k \to \infty,$$

for this particular example. Consequently (5.4) yields

$$k(\delta) \le c\delta^{-\frac{2\mu+1+n}{2\nu+1+n}/2\mu+2}.$$

Since the exact solution $x^{(\nu)}$ belongs to $\mathcal{R}(T^\mu)$ for every $\nu > \mu$, the exponent on the right-hand side can be made arbitrarily close to $-(2\mu+2)^{-1}$ by varying ν, and the proof is complete. □

The estimate of Theorem 5.1 is the best possible *uniform* bound if no further information on the spectral family $\{E_\lambda\}$ associated with T is available. It will be shown next that better bounds are possible if the spectral family has specific properties. As mentioned in the introduction, an important class of ill-posed problems are Fredholm integral equations of the first kind, where T is a compact operator. In this case E_λ is constant up to countably many jumps at eigenvalues λ_j of T. The following result shows that much fewer iterations are required for such problems.

Theorem 5.2 *Let T be a non-degenerate compact operator, and let the exact right-hand side y satisfy Assumption 3.6. Furthermore, assume that $\|y - y^\delta\| \le \delta$.*

(i) *If the eigenvalues of T decay like $O(j^{-\alpha})$ as $j \to \infty$ with some $\alpha > 0$ then*

$$k(\delta) \le c\,(\omega/\delta)^{1/(\mu+1)(\alpha+2)}. \tag{5.5}$$

(ii) *If the eigenvalues of T decay like $O(q^j)$ as $j \to \infty$ with some $q < 1$ then*

$$k(\delta) \le c\,(1 + \log^+(\omega/\delta)).$$

Here, as usual, $\log^+ t = \log t$ for $t \ge 1$ and $\log^+ t = 0$ elsewhere.

Proof. The proof is given for $n = 1$ (coinciding with the case $n = 0$) only. The extension to $n > 1$ can be done as in the proof of Theorem 5.1 and is left to the reader.

Without loss of generality assume that the eigenvalues $\{\lambda_j\}$ of T are mutually different and in decreasing order. Choose $m \in \mathbb{N}$, $0 \le m \le k := k(\delta) - 1$, and define

$$\psi_m(\lambda) = \prod_{j=1}^{m}(1 - \frac{\lambda}{\lambda_j}).$$

Let φ_{k-m} be the translated Jacobi polynomial (5.2) of degree $k - m$ with parameter $\nu = 2\mu + 2$, and set

$$p(\lambda) = \psi_m(\lambda)\varphi_{k-m}(\lambda/\lambda_{m+1}).$$

Clearly, $p \in \Pi_k^0$ with $p(\lambda_j) = 0$ for $j = 1, \ldots, m$. Furthermore, since $0 \leq \psi_m(\lambda) \leq 1$ for $0 \leq \lambda \leq \lambda_{m+1}$, (5.3) implies that $|p(\lambda)| \leq 1$ in this interval, and

$$|\lambda^{\mu+1} p(\lambda)| \leq |\lambda^{\mu+1} \varphi_{k-m}(\lambda/\lambda_{m+1})| \leq c(k - m + 1)^{-2\mu-2} \lambda_{m+1}^{\mu+1}, \qquad 0 \leq \lambda \leq \lambda_{m+1}.$$

If $\lambda_j = O(j^{-\alpha})$ as $j \to \infty$, let $m \doteq \frac{\alpha}{\alpha+2} k$ which gives

$$|\lambda^{\mu+1} p(\lambda)| \leq c(k+1)^{-(2+\alpha)(\mu+1)}, \qquad \lambda \in \sigma(T).$$

It follows that

$$\|p(T)y^\delta\| \leq \|p(T)(y - y^\delta)\| + \|p(T)T^{\mu+1}w\| \leq \delta + c(k+1)^{-(2+\alpha)(\mu+1)}\omega, \qquad (5.6)$$

and hence, Proposition 2.1 implies

$$\tau\delta \leq [p_k^{[1]}, p_k^{[1]}]_0^{1/2} \leq [p, p]_0^{1/2} \leq \delta + ck(\delta)^{-(2+\alpha)(\mu+1)}\omega.$$

This implies assertion (i).

If $\lambda_j = O(q^j)$ as $j \to \infty$, let $m = k$ in which case

$$|\lambda^{\mu+1} p(\lambda)| \leq c\tilde{q}^{k+1}, \qquad \lambda \in \sigma(T),$$

with $\tilde{q} = q^{\mu+1} < 1$, and

$$\|p(T)y^\delta\| \leq \delta + c\tilde{q}^{k+1}\omega.$$

It follows as above that

$$\tilde{q}^{k(\delta)} \geq \frac{\tau - 1}{c} \frac{\delta}{\omega},$$

which implies (ii). □

Note that Theorem 5.2 always improves upon the former bound of Theorem 5.1. The improvement can be quite dramatically. For example, if T is a Fredholm integral equation of the first kind over a bounded domain Ω with continuous kernel $h(\cdot, \cdot)$, then Mercer's theorem (cf. [71, Sect. 98]) implies

$$\sum_{j=1}^{\infty} \lambda_j = \int_\Omega h(t, t)\, dt < \infty,$$

showing that $\lambda_j = O(1/j)$, $j \to \infty$. In case of additional smoothness of h, for instance Lipschitz continuity, faster decay of the eigenvalues can be shown (cf. [70]), and conjugate gradient type methods will require (significantly) less iterations to reach order-optimal accuracy than in the general case.

Remark. The conjugate gradient type methods of Section 2.3, e.g., CGNE and CGME, can be analyzed in the same way. If Assumption 3.6 is fulfilled for some $\mu > 0$, the bound of Theorem 5.1 becomes

$$k(\delta) \le c\,(\omega/\delta)^{1/\mu+1}\,. \tag{5.7}$$

Assume next that T is a compact operator with positive *singular values* $\{\sigma_j\}$. In this case one has to replace T by TT^* in (5.6) and $\mu + 1$ by $(\mu + 1)/2$; recall that the eigenvalues $\{\lambda_j\}$ of TT^* are given by $\lambda_j = \sigma_j^2$. From this follows that assertion (ii) of Theorem 5.2 remains unchanged if the singular values of T decay geometrically. For sublinear decay rates the corresponding bound for $k(\delta)$ is

$$k(\delta) \le c\,(\omega/\delta)^{1/(\mu+1)(\alpha+1)} \qquad \text{if} \quad \sigma_j = O(j^{-\alpha})\,. \tag{5.8}$$

Again, (5.8) is always better than (5.7). Note that Fredholm integral equations of the first kind with $\mathcal{L}^2$-kernels are Hilbert-Schmidt operators, and for Hilbert-Schmidt operators one always has $\alpha \ge 1/2$. If the kernel of the integral equation is smooth, e.g., differentiable, then the singular values decay faster (cf. [5, Theorem 4.21]) and the second bound becomes even better.

When T is selfadjoint and semidefinite one can use either of the two families of conjugate gradient type methods; those of Section 2.1 *and* those of Section 2.3. Note that the algorithms of the second group require approximately twice as much work per iteration since each iteration involves one multiplication with T *and* one with T^*. Furthermore, by comparing (5.7) with (5.1) one may expect that the methods from Section 2.3 will require more iterations; in fact, for the example of Section 4.2 the stopping index $k(\delta)$ will approximately be squared when passing from a conjugate gradient type method for $Tx = y^\delta$ to a method using TT^* instead. For compact operator equations this gap becomes less dramatic when the eigenvalues decay fast to zero, as can be seen from Theorem 5.2 and the remark following it. Nevertheless, CG and MR are much more efficient than CGNE and CGME, respectively. In other words, when T is selfadjoint and semidefinite then one should avoid passing to the normal equation.

5.2 The counterexample revisited

By Proposition 2.1 the stopping index $k(\delta)$ as determined by the discrepancy principle is always minimal when $n = 1$, i.e., for MR. On the other hand, recall from the remark following Stopping Rule 4.7 that the discrepancy principle for MR and Stopping Rule 4.7 for CG determine the same stopping index $k(\delta)$ - at least in exact arithmetic. The intuitive feeling, on the other hand, is that CG achieves optimal accuracy faster than MR; thus one might try to stop CG earlier. This can be achieved by playing with τ in the definition of the stopping rules, i.e., by choosing τ somewhat larger for CG than for MR. Unfortunately, however, this intuitive feeling lacks a theoretical justification. In fact, two natural questions remain unanswered:

- Does CG always achieve optimal accuracy with less iterations than MR ?

- Under what conditions is CG's optimal accuracy better than the one of MR ?

Below, these questions will be investigated for the academic example constructed in Section 4.2. Thereby $k(\delta)$ always denotes the (identical) stopping index as determined by the two stopping rules; furthermore denote by $k_*^{CG}(\delta)$ and by $k_*^{MR}(\delta)$, respectively, the iteration indices for which CG and MR attain the best approximation of the exact solution.

Consider the problem $Tx = y$ of (4.7). Recall that the exact right-hand side y defines the inner product (4.8), and since $y \in \mathcal{R}(T)$ it is possible to define the inner products $[\cdot, \cdot]_n$ for all $n \geq -2$. Some basic asymptotics of the corresponding Jacobi polynomials have already been derived in Lemma 4.6; to compare CG and MR the dominating constants in these asymptotics are also required.

Lemma 5.3 *Let $[\cdot, \cdot]_0$ be as in (4.8). Then the following asymptotics hold for fixed $n \geq -2$ and $k \to \infty$:*

(i) $\quad [p_k^{[n+1]}, p_k^{[n+1]}]_n \doteq \frac{\Gamma^2(2\nu+n+3)}{2\nu+n+2} k^{-4\nu-2n-4}$,

(ii) $\quad [p_k^{[n+2]}, p_k^{[n+2]}]_n \doteq 2 \frac{(2\nu+n+3)\Gamma^2(2\nu+n+3)}{(2\nu+n+2)(2\nu+n+4)} k^{-4\nu-2n-4}$,

(iii) $\quad [p_k^{[n+3]}, p_k^{[n+3]}]_n \doteq 6 \frac{(2\nu+n+4)\Gamma^2(2\nu+n+4)}{(2\nu+n+2)(2\nu+n+3)(2\nu+n+5)(2\nu+n+6)} k^{-4\nu-2n-4}$.

Proof. (i) Refining (4.13), cf. [76], and inserting this into (4.12) gives

$$[p_k^{[n]}, p_k^{[n]}]_n \doteq \frac{1}{2} \Gamma^2(2\nu + n + 2) k^{-4\nu-2n-3} .$$

From (2.21) therefore follows

$$[p_k^{[n+1]}, p_k^{[n+1]}]_n \doteq \frac{4\nu+2n+4}{2} \Gamma^2(2\nu + n + 2) k^{-4\nu-2n-4} ,$$

and a final application of the functional equation $\Gamma(\zeta + 1) = \zeta\Gamma(\zeta)$ establishes (i).

(ii) As in (4.28) one concludes from the orthogonality relations that

$$\begin{aligned}
[p_k^{[n+2]}, p_k^{[n+2]}]_n &= [p_k^{[n+2]}, p_k^{[n+2]} - p_{k+1}^{[n]}]_n \\
&= (p_k^{[n+2]} - p_{k+1}^{[n]})'(0) [p_k^{[n+2]}, 1]_{n+1} \\
&= (p_k^{[n+2]} - p_{k+1}^{[n]})'(0) [p_k^{[n+2]}, p_k^{[n+2]}]_{n+1} .
\end{aligned}$$

As follows from [76, (4.21.7)],

$$p_k^{[n]\prime}(0) = -\frac{k(k + 2\nu + n + 3/2)}{2\nu + n + 2} , \tag{5.9}$$

78

hence,
$$(p_k^{[n+2]} - p_{k+1}^{[n]})'(0) \doteq \frac{2}{(2\nu+n+2)(2\nu+n+4)} k^2 .$$

Together with (i), (ii) follows.

(iii) The same technique applied to $[p_k^{[n+3]}, p_k^{[n+3]}]_n$ yields

$$
\begin{aligned}
[p_k^{[n+3]}, p_k^{[n+3]}]_n \\
= [p_k^{[n+3]}, p_k^{[n+3]} - p_{k+1}^{[n]}]_n \\
= (p_k^{[n+3]} - p_{k+1}^{[n]})'(0) [p_k^{[n+3]}, 1]_{n+1} + \frac{1}{2} (p_k^{[n+3]} - p_{k+1}^{[n]})''(0) [p_k^{[n+3]}, 1]_{n+2} \\
= (p_k^{[n+3]} - p_{k+1}^{[n]})'(0) [p_k^{[n+3]}, 1 - p_{k+1}^{[n+1]}]_{n+1} \\
\quad + \frac{1}{2} (p_k^{[n+3]} - p_{k+1}^{[n]})''(0) [p_k^{[n+3]}, p_k^{[n+3]}]_{n+2} \\
= \left(-(p_k^{[n+3]} - p_{k+1}^{[n]})'(0) \, p_{k+1}^{[n+1]'}(0) + \frac{1}{2} (p_k^{[n+3]} - p_{k+1}^{[n]})''(0) \right) [p_k^{[n+3]}, p_k^{[n+3]}]_{n+2} .
\end{aligned}
$$

From (5.9) follows

$$- (p_k^{[n+3]} - p_{k+1}^{[n]})'(0) \, p_{k+1}^{[n+1]'}(0) \doteq \frac{3}{(2\nu+n+2)(2\nu+n+3)(2\nu+n+5)} k^4 .$$

The second derivative of $p_k^{[n]}$ can be computed by similar means as in (5.9), which gives

$$p_k^{[n]''}(0) = \frac{k(k-1)(k+2\nu+n+3/2)(k+2\nu+n+5/2)}{(2\nu+n+2)(2\nu+n+3)} ,$$

hence,

$$\frac{1}{2} (p_k^{[n+3]} - p_{k+1}^{[n]})''(0) \doteq \frac{-3(2\nu+n+4)}{(2\nu+n+2)(2\nu+n+3)(2\nu+n+5)(2\nu+n+6)} k^4 .$$

Consequently,

$$[p_k^{[n+3]}, p_k^{[n+3]}]_n \doteq \frac{6}{(2\nu+n+2)(2\nu+n+3)(2\nu+n+5)(2\nu+n+6)} k^4 [p_k^{[n+3]}, p_k^{[n+3]}]_{n+2} ,$$

and the assertion follows from (i). $\qquad\square$

Consider the behavior of CG and MR for the above example given the perturbed right-hand side y^δ from (4.9). As in Section 4.3, denote by $\{x_k^\delta\}$ the iterates of CG and by $\{z_k^\delta\}$ the iterates of MR. Recall that the residual polynomials for the two methods are defined via the inner product (4.10), not via (4.8); the residual polynomials $\{p_k^\delta\}$ of CG have been computed in Proposition 4.5, namely

$$p_k^\delta = p_k^{[0]} - \vartheta_k \lambda p_{k-1}^{[2]}, \qquad \vartheta_k = \delta^2 [p_{k-1}^{[2]}, p_{k-1}^{[2]}]_1^{-1} .$$

Since the error can be rewritten as

79

$$x^{(\nu)} - x_k^\delta = p_k^\delta(T)x^{(\nu)} + p_k^{\delta\prime}(0)(y^\delta - y),$$

this gives

$$
\begin{aligned}
\|x^{(\nu)} - x_k^\delta\|^2 &= [p_k^\delta, p_k^\delta]_{-2} + (p_k^{\delta\prime}(0)\delta)^2 \\
&= [p_k^{[0]}, p_k^{[0]}]_{-2} - 2\vartheta_k [p_k^{[0]}, p_{k-1}^{[2]}]_{-1} + \vartheta_k^2 [p_{k-1}^{[2]}, p_{k-1}^{[2]}]_0 + (p_k^{[0]\prime}(0) - \vartheta_k)^2\delta^2 \, ;
\end{aligned}
$$

note that $[p_k^{[0]}, p_{k-1}^{[2]}]_{-1} = [p_k^{[0]}, p_k^{[0]}]_{-1}$ by orthogonality. Thus, substituting

$$t = \tfrac{2\nu+2}{\Gamma^2(2\nu+3)} k^{4\nu+4}\delta^2 , \tag{5.10}$$

one eventually obtains by using Lemma 5.3 and (5.9):

$$\|x^{(\nu)} - x_k^\delta\|^2 \doteq \xi_0 t^{-\nu/\nu+1}(1 + \xi_1 t + \xi_2 t^2 + \xi_3 t^3)\delta^{2\nu/\nu+1}, \qquad k \to \infty,$$

where

$$\xi_0 = (\tfrac{\Gamma^2(2\nu+3)}{2\nu+2})^{1/\nu+1} \tfrac{1}{\nu(2\nu+1)(2\nu+2)^2}, \quad \xi_1 = \tfrac{\nu(4\nu^2+8\nu+1)}{2\nu+3}, \quad \xi_2 = \tfrac{2\nu(2\nu+1)(2\nu+5)}{(2\nu+3)(2\nu+4)}, \quad \xi_3 = \tfrac{\nu(2\nu+1)}{(2\nu+3)^2} .$$

To determine the asymptotic behavior of $k_*^{CG}(\delta)$ the above expression has to be minimized, which leads to the following equation for t:

$$t^3 + (2\nu + 5)t^2 + \tfrac{4\nu^2+8\nu+1}{2\nu+1} t - \tfrac{2\nu+3}{2\nu+1} = 0 .$$

Dividing out the trivial root at $t = -1$, one readily obtains

$$t_*^{CG} \doteq -(\nu + 2) + \sqrt{(\nu+2)^2 + \tfrac{2\nu+3}{2\nu+1}}, \qquad \delta \to 0, \tag{5.11}$$

and $k_*^{CG}(\delta)$ is obtained from t_*^{CG} via (5.10).

Because of the particular form of the perturbation $y^\delta - y$, the residual polynomials of MR are the same for y and for y^δ, namely $\{p_k^{[1]}\}$. It follows as above by using (5.9) that

$$\|x^{(\nu)} - z_k^\delta\|^2 = [p_k^{[1]}, p_k^{[1]}]_{-2} + (p_k^{[1]\prime}(0)\delta)^2 \doteq \zeta_0 t^{-\nu/\nu+1}(1 + \zeta_1 t)\delta^{2\nu/\nu+1}, \qquad k \to \infty,$$

with

$$\zeta_0 = (\tfrac{\Gamma^2(2\nu+3)}{2\nu+2})^{1/\nu+1} \tfrac{3}{\nu(2\nu+1)(2\nu+3)(2\nu+4)}, \qquad \zeta_1 = \tfrac{\nu(2\nu+1)(2\nu+4)}{3(2\nu+3)} .$$

Thus, via (5.10), the optimal iteration index $k_*^{MR}(\delta)$ for MR can be obtained from

$$t_*^{MR} \doteq 3\,\frac{2\nu + 3}{(2\nu + 1)(2\nu + 4)}, \qquad \delta \to 0. \tag{5.12}$$

Comparing (5.12) and (5.11) it is easily seen that, as $\delta \to 0$,

80

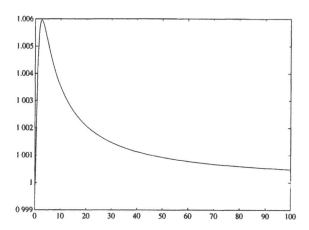

Fig. 5.1. Asymptotic ratio of optimal MR error over optimal CG error in dependence of ν

$$t_*^{CG} \doteq \frac{2\nu+3}{2\nu+1}\left(\nu+2+\sqrt{(\nu+2)^2+\tfrac{2\nu+3}{2\nu+1}}\right)^{-1} \leq \frac{2\nu+3}{(2\nu+1)(2\nu+4)} \doteq \frac{1}{3}t_*^{MR}.$$

In other words, at least asymptotically as $\delta \to 0$, the optimal iteration index of CG is smaller than the optimal iteration index of MR.

Consider Figure 5.1 for a comparison of the optimal accuracy of the two methods for $0 < \nu \leq 100$: the figure shows the limit as $\delta \to 0$ of the ratio of the optimal errors, i.e.,

$$\lim_{\delta \to 0} \|x^{(\nu)} - z^{\delta}_{k^*_{MR}(\delta)}\| / \|x^{(\nu)} - x^{\delta}_{k^*_{CG}(\delta)}\|.$$

The ratio is surprisingly close to one, i.e., the optimal errors of the two methods become almost identical as $\delta \to 0$ for this particular example. It can be shown that the limit of the plotted ratio is one, both as $\nu \to \infty$ and as $\nu \to 0$. Consequently, CG is more efficient, at least for this (academic) example: essentially the same accuracy is obtained with clearly fewer iterations.

Finally, consider the stopping index $k(\delta)$ as determined by Stopping Rules 3.10 and 4.7. Since

$$\|y^\delta - Tz^\delta_k\|^2 = [p^{[1]}_k, p^{[1]}_k]^\delta = \delta^2 + [p^{[1]}_k, p^{[1]}_k]_0 \doteq \delta^2 + \frac{\Gamma^2(2\nu+3)}{2\nu+2}k^{-4\nu-4} \qquad (5.13)$$

as $k \to \infty$, it follows that

$$k(\delta) \doteq ((\tau^2-1)^{-1}\tfrac{\Gamma^2(2\nu+3)}{2\nu+2})^{1/4\nu+4}\delta^{-1/2\nu+2}, \qquad \delta \to 0.$$

Thus, relating τ with t of (5.10) gives

$$\tau \doteq (1+\frac{1}{t})^{1/2}, \qquad \delta \to 0. \qquad (5.14)$$

81

In other words, there exist τ_*^{CG} and τ_*^{MR} such that the respective stopping indices $k(\delta)$ coincide with $k_*^{CG}(\delta)$ and $k_*^{MR}(\delta)$. Unfortunately, however, these numbers for τ are problem dependent, since they depend on the (typically unknown) parameter ν. Since $k_*^{CG}(\delta) < k_*^{MR}(\delta)$ asymptotically it is clear that $\tau_*^{CG} > \tau_*^{MR}$; in this example the ratio τ_*^{CG}/τ_*^{MR} ranges between 1.4 and 1.7 for relevant values of ν. These numbers agree with the ones that are obtained experimentally in the following section.

For comparison: the heuristic stopping rules determine the stopping index $k(y^\delta)$ as the minimum of the sequence

$$\eta_k^2 = |p_k^{[1]\prime}(0)|^2 \, [p_k^{[1]}, p_k^{[1]}]^\delta, \qquad k \geq 1,$$

for MR (cf. Stopping Rule 3.13) and

$$\eta_k^2 = |p_k^{[0]\prime}(0)|^2 \, [p_k^{[1]}, p_k^{[1]}]^\delta, \qquad k \geq 1,$$

for CG (cf. Stopping Rule 4.10), respectively. Asymptotically, these sequences only differ by the factor $(2\nu+2)^2/(2\nu+3)^2$, cf. (5.9), hence it suffices to consider only one of them. As can be seen from (5.9) and (5.13), the minimum is attained at

$$k(y^\delta) \doteq (\nu \, \tfrac{\Gamma^2(2\nu+3)}{2\nu+2})^{1/4\nu+4} \delta^{-1/2\nu+2}, \qquad \delta \to 0.$$

In view of (5.10) this corresponds to

$$l \doteq \nu, \qquad \delta \to 0. \tag{5.15}$$

In conclusion, the corresponding errors decrease with *order-optimal accuracy* even though the noise level δ is unknown. This is in agreement with Corollary 3.15, since (3.29) holds with $\gamma = 1$. For the considered stopping rules the corresponding values of l, cf. (5.14) and (5.15) do not behave like the optimal values l_*^{CG} and l_*^{MR} as ν approaches 0 or ∞: all stopping rules terminate the iteration somewhat too late when ν gets large, while the heuristic rules terminate the iteration somewhat too early when ν is very small.

5.3 An application: image reconstruction

Let $x : \mathbb{R}^2 \to \mathbb{R}_0^+$ be a two dimensional function, representing e.g. a grey-scale image. In optics one is often interested in recovering x from a blurred and noisy copy y. In many applications, blur can be modelled by a space-invariant point spread function $h : \mathbb{R}^2 \to \mathbb{R}$, in which case x and y satisfy the first kind convolution equation

$$y(s,t) = (Tx)(s,t) = \int_{\mathbb{R}^2} h(s-s', t-t') \, x(s',t') \, ds' dt'. \tag{5.16}$$

If $h \in \mathcal{L}^1(\mathbb{R}^2)$ then the spectrum of the integral operator T is a continuum given in terms of the Fourier transform of h. By the Riemann-Lebesgue lemma the spectrum of T clusters at $\lambda = 0$ showing that the deconvolution $y \mapsto x$ is ill-posed. For a practical implementation the unbounded region of integration has to be truncated to a finite domain, e.g., a square, in which case the integral operator becomes compact. Consequently the spectrum is no longer a continuum, but it is asymptotically dense in the spectrum of T.

For the reconstruction of astronomical images the point spread function h is sometimes taken to be the Gaussian

$$h(s, t) = \exp(-\chi(s^2 + t^2)), \tag{5.17}$$

in which case (5.16) is a simple model for the effect of atmosperic turbulences caused by random variations in the refractive index. This is the example which will be discussed below. The implementation of the integral equation (5.16) follows the standard approach in the literature: the image is represented by an equidistant mesh of $m \times m$ pixels with mesh-width d, and the integrals per pixel are approximated by the rectangular quadrature rule; here, $m = 64$ and $d = 1$. Note that d is just an intensive parameter as opposed to the "real" extensive mesh-width which is intrinsic in $\chi = 0.1$ and in the particular model (5.17). For practical reasons, blur is restricted to compact support, i.e., h is truncated for $\max\{|s|, |t|\} \geq 3$.

Consider Figure 5.2 for an illustration of the blur. It shows the true image and a blurred and noise corrupted copy; note that the noise (1%, see below) is almost negligible as can be seen from the flat parts of the image. The sharp contours of the exact solution make it difficult to obtain good reconstructions; fortunately, the truncation of h makes the problem better behaved – in view of Assumption 3.6 – since the Fourier transforms of the image and the truncated point spread function, respectively, have similar asymptotic behavior.

The finite dimensional linear system

$$A\mathbf{x} = \mathbf{b} \tag{5.18}$$

has a lot of structure. The "image vectors" of dimension $64 \cdot 64 = 4096$ are obtained in a straightforward way by simple row-wise ordering. Matrix A is therefore a block matrix with 64×64 blocks of dimension 64, each. Every block is Toeplitz, i.e., constant along the diagonals and only the central five diagonals are nonzero. The blocks themselves remain the same along each block-diagonal, that is, A is block-Toeplitz with Toeplitz blocks. Moreover, A is a sparse matrix with at most $5 \cdot 5 = 25$ nonzeros per row. Thus, iterative methods are highly sophisticated for solving (5.18).

To simulate noise, the exact right-hand side $\mathbf{b}$ is perturbed by normally distributed random vectors $\mathbf{e}$. For each of two noise levels, namely

$$\|\mathbf{e}\| / \|\mathbf{b}\| = 0.01, \qquad \text{resp.} \qquad \|\mathbf{e}\| / \|\mathbf{b}\| = 0.001,$$

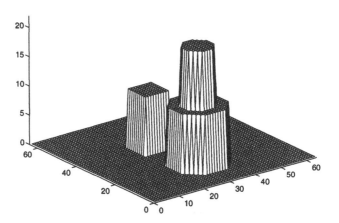

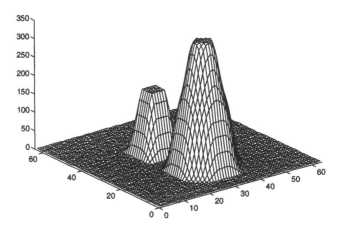

Fig. 5.2. Exact solution **x** (top) and perturbed right-hand side **b** + **e** (bottom)

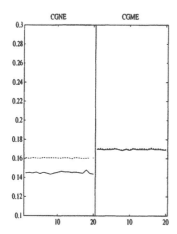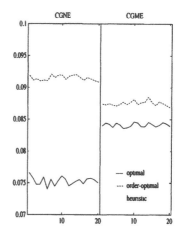

Fig. 5.3. Relative errors of the different methods for 1% noise (left) and 0.1% noise (right)

twenty perturbations are considered. Although being symmetric, A is not semidefinite, so that conjugate gradient type methods cannot be applied straight to (5.18) as in Section 2.1. Instead, approximations of $\mathbf{x}$ are computed by CGNE and CGME, using both suggested stopping rules: the order-optimal rule and the heuristic rule. Later in Section 6.6 these methods will be compared to a conjugate gradient type method that applies straight to selfadjoint indefinite ill-posed problems.

For a comparison of the accuracy of the iterative schemes, cf. Figure 5.3 and Table 5.1. They show the error of the optimal iterate (opt./solid line), the error of the iterate determined by the order-optimal stopping rule using the prior information $\delta = \|e\|$ and parameter $\tau = 1.1$ (ord./dashed line), and the error of the iterate determined by the heuristic stopping rule (heu./dotted line). In each of the two plots in Figure 5.3 (note the different scales for the two plots), the left-hand twenty columns show the errors of CGNE for the twenty noise samples, the right-hand twenty columns correspond to CGME (the same noise samples were used for the two iterative methods). The average values constitute the entries of Table 5.1.

As can be seen from these numbers, the accuracy of CGME is in general worse by about 10% as compared to CGNE. This comes somewhat unexpected in view of the results of the foregoing section; one might guess that this phenomenon is due to a higher round-off sensitivity of CGME, however, other equivalent implementations (like the bidiagonalization method of PAIGE [62]) have led to the same results; compare also Section 2.6. The order optimal rules provide satisfactory approximations, and so do the heuristic rules in the presence of less noise. In fact, for 0.1% noise the heuristic rules perform even better than the order-optimal ones; this is due to the fact that the nullspace of A is trivial, which enables a decrease of the residual norm to about $0.5\|e\|$

	1% noise			0.1% noise		
	opt.	ord.	heu.	opt.	ord.	heu.
CGNE	0.1452	0.1606	0.2514	0.0753	0.0915	0.0811
CGME	0.1695	0.1702	0.2494	0.0841	0.0875	0.0843

Table 5.1. Average (relative) error norm

	1% noise			0.1% noise		
	opt.	ord.	heu.	opt.	ord.	heu.
CGNE	26.3	13.0	2.0	166.1	68.6	96.8
CGME	12.0	13.0	2.0	92.6	68.6	86.6

Table 5.2. Average iteration count

until the optimal approximation of x is obtained; in other words, the order-optimal rules with parameter $\tau > 1$ terminate the iteration too early.

The average stopping indices are shown in Table 5.2. It can be seen that the optimal stopping index of CGME is much smaller than the one of CGNE; only about half as many iterations are required for CGME (but note that the accuracy is somewhat worse). Recall that a similar observation could be established theoretically for the example considered in Section 5.2.

The numbers in Table 5.2 also explain the loss of accuracy of the heuristic stopping rule in case of 1% noise: the stopping index $k = 2$ is much too small. As mentioned before, the performance is much better for 0.1% noise; for particular noise samples the heuristic stopping index for CGME and the optimal one even coincide. The above observation is in agreement with the theoretical results obtained for the example in Section 5.2. There it has been shown that the heuristic stopping rules stop the iteration somewhat too early if Assumption 3.6 is fulfilled for smaller μ only. This is precisely the case in the present application. As δ becomes small, however, the good performance of the heuristic rule can be verified by means of the theoretical a posteriori bounds obtained in Theorems 3.14 and 4.11: in the case of 0.1% noise the estimates δ_* for the actual noise level have always been in the interval

$$\delta_*/\|\mathbf{b}\| \ \in \ [0.0007, 0.00085] \,.$$

Thus, the ratio $\|y - y^\delta\|/\delta_*$ which enters the aforementioned bounds has always been below 1.5.

Figure 5.4 illustrates the semiconvergence of the iteration by showing the iteration histories of the (relative) errors of CGNE and CGME for two right-hand sides with 1%

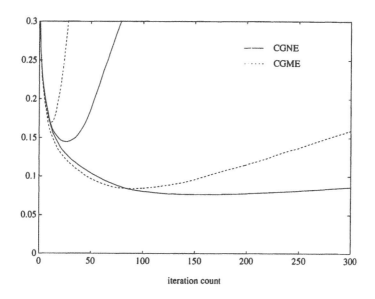

Fig. 5.4. Iteration error history for two noise levels

and 0.1% noise. Note that the two curves corresponding to the same method overlap until the first of them starts to diverge. This illustrates that the converging behavior in the beginning of the iteration is essentially independent of noise; less noise leads to delayed divergence, and hence to better accuracy.

Figure 5.5 shows the error estimate for CGNE provided by the heuristic Stopping Rule 3.13. The corresponding plot for CGME is shown in Figure 5.6. In this particular case the stopping index for CGME coincides with the optimal iteration error, whereas the stopping index for CGNE is too small. It can nicely be seen that the dashed lines clearly underestimate the converging component of the error in the beginning of the iteration; the diverging components are matched reasonably well. This again explains the rapid termination according to the heuristic rules for larger perturbations of the right-hand side, cf. Table 5.2. Note, however, that it is obviously better to stop the iteration too early than to stop too late, because in the latter case δ_* may become very small and the error bounds of Theorems 3.14 and 4.11 may blow up.

Finally, see Figure 5.7 for a plot of the corresponding reconstructions obtained from CGME and CGNE with the heuristic stopping rules. The quality of the results is comparable. The important contours of the true image, cf. Figure 5.2, are recovered.

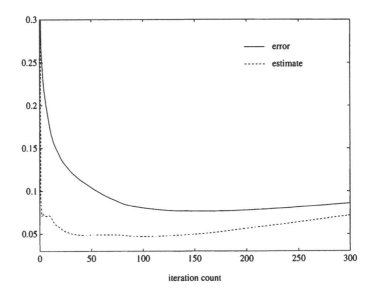

Fig. 5.5. Iteration error of CGNE and heuristic error estimate

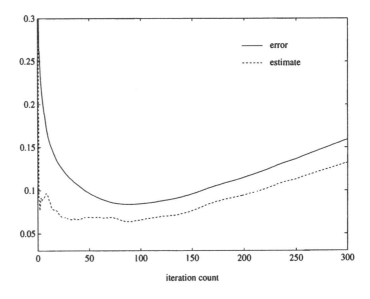

Fig. 5.6. Iteration error of CGME and heuristic error estimate

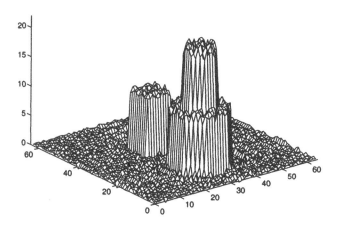

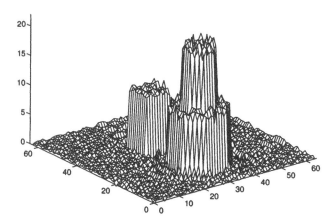

Fig. 5.7. Approximations obtained by CGNE (top) and CGME (bottom)

Notes and remarks

Section 5.1. Theorem 5.1 is due to NEMIROVSKII [58], see also BRAKHAGE [6]. The estimate (5.5) in Theorem 5.2 is taken from NEMIROVSKII and POLYAK [59]; compare LOUIS [53] for a suboptimal bound. Examples can be constructed to show that the exponent in (5.5) cannot be improved in general, cf. [36].

Section 5.3. The example of this section is borrowed from [39]; for more numerical examples with this test problem, see [38, Section 8.2]. An elementary treatment of image reconstruction and image restoration can be found in the book by LAGENDIJK and BIEMOND [49]; the reference to the particular point spread function (5.17) is on p. 30.

6. A Minimal Residual Method for Indefinite Problems

This final chapter considers the applicability of Krylov subspace methods to selfadjoint, indefinite problems. There are a number of important applications where it pays to avoid forming the positive (semi)definite normal equation; two such applications are described in Sections 6.6 and 6.7. A family of conjugate gradient type methods, suitable for ill-posed indefinite problems, is introduced in Section 6.1. These methods have very similar optimality properties as the methods of the foregoing chapters. In particular, one of these schemes is a minimal residual type method, and this one will be analyzed exemplarily for its convergence and regularizing properties. The main result (in Section 6.4) states that the discrepancy principle provides an order-optimal stopping rule.

6.1 MR-II, a variant of MR

The conjugate gradient type methods of the previous chapters are defined by residual polynomials $p_k \in \Pi_k^0$, which form a system of orthogonal polynomials with respect to the inner product

$$[\varphi, \psi]_n = \langle \varphi(T)(y - Tx_0), T^n \psi(T)(y - Tx_0) \rangle.$$

If T is a selfadjoint but indefinite operator, the above expression can be rewritten as

$$[\varphi, \psi]_n = \int_{-\infty}^{\infty} \varphi(\lambda)\psi(\lambda)\,\lambda^n d\|E_\lambda(y - Tx_0)\|^2,$$

showing that the polynomials $\{p_k\}$ are orthogonal with respect to some measure supported on the positive *and* on the negative semiaxis. Note, however, that the inner product $[\cdot, \cdot]_n$ is only definite when $n \in \mathbb{N}_0$ is even. Even then it may happen (and if $d\|E_\lambda(y - Tx_0)\|^2$ is symmetric with respect to the origin it *will* happen for every odd k) that a zero of the orthogonal polynomial comes to lie in the origin. Of course, this implies that no multiple of the corresponding polynomial belongs to Π_k^0, and hence, the original definition of conjugate gradient type methods may fail for indefinite problems.

Alternatively, conjugate gradient type methods for indefinite problems can be defined by the optimality property of Proposition 2.1. According to this condition, x_k is taken to satisfy

$$\|T^{n/2}(y - Tx_k)\| \leq \|T^{n/2}(y - Tx)\| \tag{6.1}$$

for all $x \in x_0 + \mathcal{K}_{k-1}(y - Tx_0; T)$. Again, for indefinite T, (6.1) is not useful for all $n \in \mathbb{N}_0$, only for n even. In this case the solutions of (6.1) for $k = 1, 2, \ldots$, can be computed by short recurrences (compare the remark before Lemma 6.4), but their regularizing properties are not understood.

For the purpose of regularization it seems to be more appropriate to search for the minimizer of (6.1) within the Krylov subspace

$$x_0 + \mathcal{K}_{k-2}(T(y - Tx_0); T) \tag{6.2}$$

originating from $T(y - Tx_0)$. The corresponding iterate x_k can still be written as in (2.1), namely

$$x_k = x_0 + q_{k-1}(T)(y - Tx_0), \tag{6.3}$$

but the iteration polynomial q_{k-1} now has a root at the origin. Therefore, the corresponding residual polynomial p_k satisfies a Hermite interpolation condition, i.e.,

$$p_k \in \Pi_k^{00} = \{p \in \Pi_k \mid p(0) = 1, \; p'(0) = 0\}. \tag{6.4}$$

Note that x_k minimizes (6.1) in the Krylov space (6.2), if and only if its residual polynomial p_k satisfies

$$[p_k, p_k]_n \leq [p, p]_n \qquad \text{for all } p \in \Pi_k^{00}. \tag{6.5}$$

The following result characterizes the minimizer of (6.5) in terms of orthogonality relations; an efficient numerical scheme for the computation of $\{x_k\}$ on the basis of these orthogonality relations is derived afterwards. Recall that κ always denotes the number of nonzero points of increase of $\|E_\lambda(y - Tx_0)\|^2$, including the case when $\kappa = \infty$.

Proposition 6.1 *Let $n \in \mathbb{N}_0$ be even, $0 \leq k \leq \kappa + 1$. There is a unique $p_k \in \Pi_k^{00}$ minimizing (6.5), and this polynomial is characterized by the following relation:*

$$[p_k, p]_{n+2} = 0 \qquad \text{for all } p \in \Pi_{k-2}. \tag{6.6}$$

Proof. Any $p \in \Pi_k^{00}$ can be rewritten as $p = 1 - \lambda^2 s$ with some polynomial $s \in \Pi_{k-2}$, and vice versa. Since Π_{k-2} is a Hilbert space with inner product $[\cdot, \cdot]_n$ as long as $k - 2 < \kappa$, the minimization problem

$$[p, p]_n = [1 - \lambda^2 s, 1 - \lambda^2 s]_n \longrightarrow \min$$

has a unique solution $s_{k-2} \in \Pi_{k-2}$ corresponding to a unique solution $p_k \in \Pi_k^{00}$ of (6.5).

92

Let $p_k \in \Pi_k^{00}$, and let p be any polynomial in Π_{k-2}. It follows that $p_k - \gamma\lambda^2 p \in \Pi_k^{00}$ for every $\gamma \in \mathbf{R}$, and

$$[p_k - \gamma\lambda^2 p, p_k - \gamma\lambda^2 p]_n - [p_k, p_k]_n = \gamma^2 [p, p]_{n+4} - 2\gamma[p_k, p]_{n+2}.$$

If $[p_k, p]_{n+2} \neq 0$ then the right-hand side can be made negative by choosing γ appropriately. It therefore follows that p_k is the minimizer of (6.5) in Π_k^{00}, if and only if (6.6) holds. □

If $\kappa < \infty$ then the polynomial $p_{\kappa+1}$ is the unique polynomial in $\Pi_{\kappa+1}^{00}$ with κ roots in the nonzero points of increase of $d\|E_\lambda(y - Tx_0)\|^2$, and hence

$$[p_{\kappa+1}, p_{\kappa+1}]_n = \|(I - P)T^n y\|^2 = \begin{cases} 0, & n > 0, \\ \|(I - P)y\|^2, & n = 0. \end{cases} \tag{6.7}$$

As mentioned above, orthogonal polynomials corresponding to $[\cdot, \cdot]_n$ may have a root at $\lambda = 0$, in which case the corresponding polynomial $p_k^{[n]}$ is not well-defined. Always well-defined are the *orthonormal polynomials* $\{u_k^{[n]}\}$ with positive leading coefficient,

$$u_k^{[n]} \in \Pi_k \qquad \text{and} \qquad [u_k^{[n]}, u_j^{[n]}]_n = \delta_{jk}. \tag{6.8}$$

The following corollary of Proposition 6.1 is the basis for an efficient numerical scheme for computing the sequence $\{x_k\}$. Note the similarity to Proposition 2.5.

Corollary 6.2 *Let $n \in \mathbf{N}_0$ be even. For $1 \leq k \leq \kappa$, the minimizers p_k and p_{k+1} of (6.5) in Π_k^{00} and Π_{k+1}^{00}, respectively, are related by*

$$\frac{p_k - p_{k+1}}{\lambda^2} = \varrho_k u_{k-1}^{[n+4]}, \qquad \varrho_k = [p_k, u_{k-1}^{[n+4]}]_{n+2}. \tag{6.9}$$

Proof. Let $p := (p_k - p_{k+1})/\lambda^2$. Because of the Hermite interpolation conditions (6.4) p belongs to Π_{k-1}. By Proposition 6.1,

$$[p, q]_{n+4} = [p_k, q]_{n+2} - [p_{k+1}, q]_{n+2} = 0$$

for all $q \in \Pi_{k-2}$. Hence, p is a multiple of $u_{k-1}^{[n+4]}$, and $p_{k+1} = p_k - \varrho\lambda^2 u_{k-1}^{[n+4]}$ for some $\varrho \in \mathbf{R}$. The value of ϱ can be determined from the optimality property (6.5) of p_{k+1}. Writing

$$[p_k - \varrho\lambda^2 u_{k-1}^{[n+4]}, p_k - \varrho\lambda^2 u_{k-1}^{[n+4]}]_n = [p_k, p_k]_n - 2\varrho[p_k, u_{k-1}^{[n+4]}]_{n+2} + \varrho^2 [u_{k-1}^{[n+4]}, u_{k-1}^{[n+4]}]_{n+4},$$

it is easily seen that the right-hand side becomes minimal for $\varrho = \varrho_k$. Thus, (6.9) follows from (6.5). □

Note that ϱ_k also depends on n; since n will be set to zero in the following sections this parameter will be omitted, though.

Corollary 6.3 *For $n \in \mathbb{N}_0$ even, and $0 \le k \le \kappa$, the iterates x_k and x_{k+1} of (6.3) coincide, if and only if $p_k = p_k^{[n+2]}$, i.e., if and only if $p_k^{[n+2]}$ exists and $p_k^{[n+2]\prime}(0) = 0$.*

Proof. By virtue of (6.9) $p_k = p_{k+1}$ holds, if and only if $\varrho_k = 0$, namely if and only if

$$[p_k, u_{k-1}^{[n+4]}]_{n+2} = 0.$$

Since $u_{k-1}^{[n+4]}$ is a polynomial of exact degree $k - 1$, the above, together with Proposition 6.1, implies that $[p_k, p]_{n+2} = 0$ for all $p \in \Pi_{k-1}$. In other words, $p_k^{[n+2]}$ exists and coincides with p_k. This implies that $p_k^{[n+2]\prime}(0) = 0$. Vice versa, if $p_k^{[n+2]\prime}(0) = 0$ then one has $p_k^{[n+2]} \in \Pi_k^{00}$. Since $p_k^{[n+2]}$ satisfies (6.6), it equals p_k by virtue of Proposition 6.1. $\qquad\square$

The polynomials $\{u_k^{[n+4]}\}$ can be determined from their three-term recurrence formula, which has the form

$$\beta_0 = [1, 1]_{n+4}^{1/2}, \qquad u_{-1}^{[n+4]} = 0, \qquad u_0^{[n+4]} = 1/\beta_0,$$

$$\tilde{u}_{k+1}^{[n+4]} = \lambda u_k^{[n+4]} - \alpha_k u_k^{[n+4]} - \beta_k u_{k-1}^{[n+4]}, \qquad\qquad\qquad (6.10)$$

$$\qquad\qquad\qquad\qquad\qquad\qquad\qquad\qquad\qquad k \ge 0,$$

$$\beta_{k+1} = [\tilde{u}_{k+1}^{[n+4]}, \tilde{u}_{k+1}^{[n+4]}]_{n+4}^{1/2}, \qquad u_{k+1}^{[n+4]} = \tilde{u}_{k+1}^{[n+4]}/\beta_{k+1},$$

where

$$\alpha_k = [u_k^{[n+4]}, u_k^{[n+4]}]_{n+5}, \qquad k \ge 0.$$

This is easily checked using the orthonormality of $\{u_k^{[n+4]}\}$.

Once $u_{k-1}^{[n+4]}$ is known, ϱ_k can be computed from (6.9). Note that (6.9) implies that the iteration polynomials $\{q_k\}$ enjoy the update formula

$$q_k = q_{k-1} + \varrho_k \lambda u_{k-1}^{[n+4]}, \qquad k \ge 1,$$

which defines the iterates x_k via (6.3). Here, $x_1 = x_0$ since $\Pi_0^{00} = \Pi_1^{00} = \{p \equiv 1\}$. If $\kappa < \infty$ then the algorithm terminates with $\beta_\kappa = 0$, i.e., with last iterate $x_{\kappa+1}$.

The implementation of this scheme is exemplified in Algorithm 6.1 for the important case $n = 0$. Introducing the intermediate quantities

$$v_k = T u_{k-1}^{[4]}(T)(y - Tx_0), \qquad w_k = T v_k, \qquad k \ge 0,$$

and providing extra storage for $T w_{k-1}$, only one matrix vector multiply is required per iteration. Note that the computation of $\{v_k\}$ via (6.10) is nothing but the standard Lanczos process, cf. [26].

$$r_0 = y - Tx_0$$
$$x_1 = x_0$$
$$r_1 = r_0$$
$$v_{-1} = 0$$
$$v_0 = Tr_0$$
$$w_{-1} = 0$$
$$w_0 = Tv_0$$
$$\beta = \|w_0\|$$
$$v_0 = v_0/\beta$$
$$w_0 = w_0/\beta$$
$$k = 1$$
`while (not stop) do`
$$\varrho = \langle r_k, w_{k-1} \rangle$$
$$x_{k+1} = x_k + \varrho v_{k-1}$$
$$r_{k+1} = r_k - \varrho w_{k-1}$$
$$\alpha = \langle w_{k-1}, Tw_{k-1} \rangle$$
$$v_k = w_{k-1} - \alpha v_{k-1} - \beta v_{k-2}$$
$$w_k = Tw_{k-1} - \alpha w_{k-1} - \beta w_{k-2}$$
$$\beta = \|w_k\|$$
$$v_k = v_k/\beta$$
$$w_k = w_k/\beta$$
$$k = k + 1$$
`end while.`

Algorithm 6.1: MR-II

Recall that for $n = 0$ the iterate x_k minimizes the residual norm in the Krylov subspace (6.2). Hence, like the corresponding method of Section 2.2, this is a *minimal residual method*. Note, however, that in general the iterates differ from those of MR, since the residuals are minimized in different Krylov subspaces. The method of Algorithm 6.1 is therefore called MR-II further on.

Remark. At this point it is worth noting that the extension of the original MR method of the foregoing chapters (where the residual polynomial $\tilde{p}_k$ minimizes $[p, p]_n$, with $n \in \mathbf{N}_0$ even, over Π_k^0) to indefinite problems can be derived similarly. The analog of (6.9) for the difference of $\tilde{p}_{k+1}$ and $\tilde{p}_k$ is

$$\tilde{p}_{k+1} = \tilde{p}_k + \tilde{\varrho}_k \lambda u_k^{[n+2]},$$

where $\tilde{\varrho}_k$ is easily determined with the same argument as in Corollary 6.2. The result is equivalent to Proposition 2.5 (note the different meaning of n there) except for the use of orthonormal polynomials to avoid breakdowns. This MR implementation (called ORTHODIR in the literature) is fairly similar to Algorithm 6.1. In fact, the only difference is the initialization of v_0 which should be $v_0 = r_0$, cf. [7]. As said before, the reason that this algorithm is not considered in more detail here is the lack of theoretical results concerning its regularizing properties.

On the other hand, the regularizing properties of the conjugate gradient type methods acting in the modified Krylov spaces (6.2) can be analyzed completely. For ease of notational simplicity this will only be exemplified for the MR-II method, i.e., the case $n = 0$. In particular, in Section 6.4 it will be shown that the discrepancy principle provides order-optimal accuracy for the iterates of MR-II.

This analysis requires the following lemma that may be seen as an analog of Corollary 2.6.

Lemma 6.4 *Let $n \in \mathbf{N}_0$ be even, $0 \leq k \leq \kappa$. Then the following holds:*

$$|p_k''(0) - p_{k+1}''(0)| = 2\,|u_{k-1}^{[n+4]}(0)|\,([p_k, p_k]_n - [p_{k+1}, p_{k+1}]_n)^{1/2}.$$

Proof. Starting from

$$[p_k, p_k]_n - [p_{k+1}, p_{k+1}]_n = [p_k - p_{k+1}, p_k - p_{k+1}]_n + 2[p_{k+1}, p_k - p_{k+1}]_n,$$

it follows from Proposition 6.1 that the second term on the right-hand side vanishes because $p_k - p_{k+1} = \lambda^2 s_{k-1}$ with $s_{k-1} = \varrho_k u_{k-1}^{[n+4]} \in \Pi_{k-1}$, cf. (6.9). It follows that

$$[p_k, p_k]_n - [p_{k+1}, p_{k+1}]_n = [s_{k-1}, s_{k-1}]_{n+4} = \varrho_k^2 [u_{k-1}^{[n+4]}, u_{k-1}^{[n+4]}]_{n+4} = \varrho_k^2.$$

By letting $\lambda \to 0$ in (6.9) one observes that

$$p_k''(0) - p_{k+1}''(0) = 2\varrho_k u_{k-1}^{[n+4]}(0).$$ (6.11)

Combining these two equations, the claim follows. □

6.2 On the zeros of the residual polynomials

A major tool for the analysis of the conjugate gradient type methods of the previous chapters have been the many well-known properties of the zeros of the corresponding residual polynomials. Since these residual polynomials form a system of real orthogonal polynomials, all their zeros are real, and zeros of consecutive polynomials interlace.

In general, there is no such interlacing property for the zeros of the polynomials $\{p_k\}$ characterized in Proposition 6.1. Partial interlacing results are available, though, which are strong enough to extend the analysis of the foregoing sections to MR-II. These results imply, in particular, that all zeros of p_k are real with at least $k-1$ of them belonging to the convex hull of the support of the weight function.

The notion of interlacing points will be used in a less strict sense in the sequel, including the case where some points coincide. Two finite sets $\{\xi_j\}$ and $\{\eta_j\}$ are said to *interlace*, if the numbers of elements differ by at most one, and if each closed interval $[\xi_j, \xi_{j+1}]$ between two neighboring points from $\{\xi_j\}$ contains at least one point from $\{\eta_j\}$, and vice versa. Consequently, the corresponding open interval (ξ_j, ξ_{j+1}) cannot contain more than one point from $\{\eta_j\}$, but it may contain none. The closed interval may contain up to three points from $\{\eta_j\}$, in which case two of these points coincide with ξ_j and ξ_{j+1}, respectively.

Proposition 6.5 *For $k \leq \kappa$ the zeros of p_k are real and simple, and interlace with the zeros of $u_k^{[2]}$.*

Proof. Expanding p_k in terms of the orthonormal system $\{u_k^{[2]}\}$ gives

$$p_k = \sum_{j=0}^{k}[p_k, u_j^{[2]}]_2\, u_j^{[2]}.$$

By Proposition 6.1, the first $k-2$ expansion coefficients vanish, and hence, p_k is a linear combination of $u_{k-1}^{[2]}$ and $u_k^{[2]}$, i.e.,

$$p_k = \rho_k u_{k-1}^{[2]} + \sigma_k u_k^{[2]}.$$

(There is no relation between ρ_k and ϱ_k defined in Corollary 6.2). If $\rho_k = 0$ then $p_k^{[2]}$ exists and $p_k = p_k^{[2]}$. Hence, the roots of p_k and $u_k^{[2]}$ are the same in this case. Otherwise, p_k and $\rho_k u_{k-1}^{[2]}$ share signs at the roots of $u_k^{[2]}$. Because of the classical interlacing property for the zeros of orthogonal polynomials, the sign pattern of $u_{k-1}^{[2]}$

(and thus the sign pattern of p_k) at these points is alternating. Consequently, p_k must have a zero between two consecutive roots of $u_k^{[2]}$, and another zero either to the left or to the right of all the roots of $u_k^{[2]}$. □

For an analysis of the regularizing properties of MR-II it will be necessary to relate the roots of p_k with those of $u_{k-1}^{[4]}$. The link between these two sets are the zeros of $u_k^{[2]}$.

Proposition 6.6 *For $k \leq \kappa$ the roots of $u_{k-1}^{[4]}$ interlace with the roots of $u_k^{[2]}$.*

Proof. For $k = \kappa$ the assertion is the classical interlacing theorem for the zeros of orthogonal polynomials, since $u_\kappa^{[2]} = u_\kappa^{[4]}$. For $k < \kappa$ the polynomial $\lambda^2 u_{k-1}^{[4]}$ can be expanded with respect to the orthonormal basis $\{u_k^{[2]}\}$, i.e.,

$$\lambda^2 u_{k-1}^{[4]} = \sum_{j=0}^{k+1} [u_{k-1}^{[4]}, u_j^{[2]}]_4 \, u_j^{[2]} ,$$

and the first $k - 2$ expansion coefficients vanish because of orthogonality. From the three-term recurrence relation one has, compare (6.10),

$$\tilde{\beta}_{k+1} u_{k+1}^{[2]} = \lambda u_k^{[2]} - \tilde{\alpha}_k u_k^{[2]} - \tilde{\beta}_k u_{k-1}^{[2]} ,$$

for some $\tilde{\alpha}_k, \tilde{\beta}_k, \tilde{\beta}_{k+1} \in \mathbb{R}$ with $\tilde{\beta}_{k+1} \neq 0$, and hence, $u_{k+1}^{[2]}$ can be eliminated from the above expansion. Thus, there are numbers ρ_k, σ_k and τ_k (in general different from those in the proof above) with

$$\lambda^2 u_{k-1}^{[4]} = \rho_k u_{k-1}^{[2]} + (\sigma_k + \tau_k \lambda) u_k^{[2]} . \tag{6.12}$$

Two different cases must be considered. First, if $u_k^{[2]}(0) \neq 0$ then $\lambda^2 u_{k-1}^{[4]}$ and $\rho_k u_{k-1}^{[2]}$ share signs at zeros of $u_k^{[2]}$. Since λ^2 is always positive at these points, and since the zeros of $u_{k-1}^{[2]}$ and $u_k^{[2]}$ interlace, this implies that the zeros of $u_{k-1}^{[4]}$ also interlace those of $u_k^{[2]}$. Second, if $u_k^{[2]}(0) = 0$ then both ρ_k and σ_k must be zero, since the left-hand side of (6.12) has (at least) a double zero at $\lambda = 0$. Consequently, in this case the zeros of $u_{k-1}^{[4]}$ are precisely the zeros of $u_k^{[2]}$ different from $\lambda = 0$. □

The following lemma, which will play a central role in Section 6.4, provides an application of these interlacing properties. To state it properly, a few more notations are required.

As in the previous chapters the roots of p_k are denoted by $\{\lambda_{j,k}\}$, and they are assumed to be in increasing order. Since $p_k \in \Pi_k^{00}$ the reciprocals of these roots add up to zero, i.e.,

98

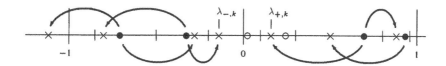

Fig. 6.1. Roots of the three polynomials p_k, $u_k^{[2]}$, and $u_{k-1}^{[4]}$

$$0 = p_k'(0) = -\sum_{j=1}^{k} \frac{1}{\lambda_{j,k}}, \qquad k \geq 2. \qquad (6.13)$$

Therefore there is at least one positive and one negative root of p_k for $k \geq 2$. The smallest positive (largest negative) root of p_k will be denoted by $\lambda_{+,k}$ ($\lambda_{-,k}$).

Throughout, let $\{\mu_{j,k-1}\}_{j=1}^{k-1}$ be the roots of $u_{k-1}^{[4]}$ with the same increasing ordering. For $k \leq \kappa$ there is an exceptional set $\mathcal{E}_{k-1}$ of up to two of these roots, which plays a special role because these zeros may come arbitrarily close to the origin in the course of the iteration.

$\mathcal{E}_{k-1}$ is specified as follows. When p_k and $u_k^{[2]}$ coincide, there is precisely one zero $\mu_{j,k-1}$ that lies strictly between $\lambda_{-,k}$ and $\lambda_{+,k}$ by Proposition 6.6. This one element forms the exceptional set $\mathcal{E}_{k-1}$ in this first case. In the other case the roots of p_k and $u_k^{[2]}$ have no points in common. By Proposition 6.5, the two roots $\lambda_{-,k}$ and $\lambda_{+,k}$ interlace with three consecutive zeros of $u_k^{[2]}$, which in turn interlace with precisely two of the roots $\{\mu_{j,k-1}\}$, as follows readily from a closer look at the proof of Proposition 6.6. These two roots form the set $\mathcal{E}_{k-1}$ in this second case. Note that this gives no information about whether both elements, just one element, or no element of $\mathcal{E}_{k-1}$ belong to $[\lambda_{-,k}, \lambda_{+,k}]$. On the other hand, every zero $\mu_{j,k-1} \in [\lambda_{-,k}, \lambda_{+,k}]$ necessarily belongs to $\mathcal{E}_{k-1}$. With this exceptional set $\mathcal{E}_{k-1}$, a prime in

$$\sum_{j}' \qquad (6.14)$$

means that the index range for j excludes indices for which $\mu_{j,k-1} \in \mathcal{E}_{k-1}$.

Compare Figure 6.1 for an illustration of the general situation with $k = 7$. In this figure a root of p_k, $u_k^{[2]}$ and $u_{k-1}^{[4]}$ is marked by a cross, a small vertical bar, and a bullet or a circle, respectively. $\lambda_{-,k}$ and $\lambda_{+,k}$ are labelled. Note that one of the roots of p_k is outside the interval $[-1, 1]$. The exceptional set $\mathcal{E}_6$ has two zeros, shown by circles rather than bullets; note that just one of these circles lies between $\lambda_{-,k}$ and $\lambda_{+,k}$. In this plot the zeros of p_k and those of $u_{k-1}^{[4]}$ do not interlace, but the general rule says that between any two consecutive zeros of the one polynomial there are at most two zeros of the other polynomial.

Lemma 6.7 *With the above notations the following holds:*

$$\left|\sum_j{}' \frac{1}{\mu_{j,k-1}}\right| \le 2\max\left\{\frac{1}{\lambda_{+,k}}, \frac{1}{|\lambda_{-,k}|}\right\}, \qquad 2 \le k \le \kappa. \tag{6.15}$$

Proof. The proof is understood most easily by considering the pointers in Figure 6.1. The technical details will be omitted. Comparing the zeros $\mu_{j,k-1} \notin \mathcal{E}_{k-1}$ of $u_k^{[4]}$ (the bullets) with those of p_k (the crosses) as indicated by the pointers *underneath* the axis one obtains

$$\begin{aligned}
\sum_{\mu_{j,k}<0}{}' \frac{1}{\mu_{j,k-1}} &\ge \sum_{\lambda_{j,k}<0} \frac{1}{\lambda_{j,k}}, \\
\sum_{\mu_{j,k}>0}{}' \frac{1}{\mu_{j,k-1}} &\le \sum_{\lambda_{j,k}>0} \frac{1}{\lambda_{j,k}}.
\end{aligned} \tag{6.16}$$

Comparing zeros as indicated by the pointers *above* the axis one obtains similarly

$$\begin{aligned}
\sum_{\mu_{j,k}<0}{}' \frac{1}{\mu_{j,k-1}} &\le \sum_{\lambda_{j,k}<0} \frac{1}{\lambda_{j,k}} - \frac{2}{\lambda_{-,k}}, \\
\sum_{\mu_{j,k}>0}{}' \frac{1}{\mu_{j,k-1}} &\ge \sum_{\lambda_{j,k}>0} \frac{1}{\lambda_{j,k}} - \frac{2}{\lambda_{+,k}}.
\end{aligned} \tag{6.17}$$

The extra terms $-2\lambda_{-,k}^{-1}$ in (6.17) are correction terms to cope with the fact that no pointers exist to one of the four roots of p_k that are closest to the origin. Adding up the inequalities in (6.16) and (6.17) one obtains the inequality chain

$$-\frac{2}{\lambda_{+,k}} + \sum_{j=1}^{k} \frac{1}{\lambda_{j,k}} \le \sum_j{}' \frac{1}{\mu_{j,k-1}} \le \sum_{j=1}^{k} \frac{1}{\lambda_{j,k}} - \frac{2}{\lambda_{-,k}}.$$

Thus the assertion follows from (6.13). □

6.3 Convergence and divergence

In this section the convergence properties of MR-II will be investigated. The analysis will be restricted to the case $x_0 = 0$. It is shown that the iterates x_k converge to $T^\dagger y$ whenever $y \in \mathcal{D}(T^\dagger)$. This slight difference ($\mathcal{D}(T^\dagger)$ instead of $\mathcal{R}(T)$) from the results of Theorems 3.4 and 3.5 is due to the additional property $p_k'(0) = 0$.

The basic ideas for proving convergence of MR-II are similar to those used in Section 3.2 to establish bounds for the norms of the residuals. However, the technical details are now significantly harder. Obviously $|p_k'(0)|$ can no longer take its pronounced role as convergence modulus from the former chapters, since it is now zero by construction. Its role will be taken over by the numbers $|p_k''(0)|^{1/2}$; note that

$$p_k''(0) = \sum_{i \neq j} \frac{1}{\lambda_{i,k} \lambda_{j,k}} = \frac{1}{2} \left(\sum_{j=1}^{k} \frac{1}{\lambda_{j,k}} \right)^2 - \frac{1}{2} \sum_{j=1}^{k} \frac{1}{\lambda_{j,k}^2} ,$$

and hence, by virtue of (6.13),

$$|p_k''(0)| = \frac{1}{2} \sum_{j=1}^{k} \frac{1}{\lambda_{j,k}^2} . \qquad (6.18)$$

Recall from the theory of orthogonal polynomials that all zeros of $u_k^{[2]}$ are contained in $[-\|T\|, \|T\|] \subset [-1, 1]$. By Proposition 6.5 this implies that at least $k - 1$ zeros of p_k belong to $[-1, 1]$ showing that

$$|p_k''(0)| \geq \frac{1}{2}(k - 1), \qquad (6.19)$$

cf. (6.18). It follows that $|p_k''(0)| \to \infty$ as $k \to \infty$, although the sequence $\{|p_k''(0)|\}$ need not be increasing, cf. Example 6.14.

Lemma 6.8 *Let $p \in \Pi_k^{00}$, $k \geq 2$, have k real (not necessarily simple) zeros $\{\lambda_j\}_{j=1}^k$. Denote by λ_- and λ_+ the negative and positive roots, respectively, which are closest to the origin. Then, for every $\nu > 0$ there is some constant $c > 0$ such that*

$$|\lambda|^{2\nu} p(\lambda) \leq c|p''(0)|^{-\nu}, \qquad \lambda_- \leq \lambda \leq \lambda_+ .$$

The constant c is independent of the polynomial and its degree.

Proof. Since all roots of p are real, p' has precisely one zero in (λ_-, λ_+), namely the one in the origin. Consequently, in this interval p only attains values between zero and one. In the same interval the maximum of $|\lambda|^{2\nu} p$ is attained at a zero $\lambda = \lambda_*$ of $(\lambda^{2\nu} p)'$. Some straightforward computations show that λ_* solves the equation

$$\frac{2\nu}{\lambda^2} = -\frac{p'(\lambda)}{\lambda p(\lambda)} . \qquad (6.20)$$

Substituting $p(\lambda) = \prod_{j=1}^{k}(1 - \lambda/\lambda_j)$, the right-hand side of (6.20) can be rewritten as

$$-\frac{p'(\lambda)}{\lambda p(\lambda)} = \sum_{j=1}^{k} \frac{-1}{\lambda(\lambda - \lambda_j)} = \sum_{j=1}^{k} \frac{1}{\lambda_j} \left(\frac{1}{\lambda} - \frac{1}{\lambda - \lambda_j} \right) = \sum_{j=1}^{k} \frac{1}{\lambda_j} \frac{1}{\lambda_j - \lambda} ,$$

because, as in (6.13), the reciprocals of λ_j sum up to zero. Inserting this into (6.20) yields the following equation for λ_*:

$$\frac{2\nu}{\lambda_*^2} = \sum_{j=1}^{k} \frac{1}{\lambda_j} \frac{1}{\lambda_j - \lambda_*} . \qquad (6.21)$$

Note that each term on the right-hand side of (6.21) is positive since $\lambda_* \in (\lambda_-, \lambda_+)$.
    Next, let

$$\varepsilon := |\lambda_*| |2p''(0)|^{1/2}. \tag{6.22}$$

The aim of the following estimations is to show that ε can be bounded from above by some constant. It follows immediately from (6.18) (with p_k replaced by p) that

$$|\lambda_*| = \varepsilon \left(\sum_{j=1}^{k} \frac{1}{\lambda_{j,k}^2}\right)^{-1/2} \le \varepsilon |\lambda_j| \qquad \text{for every } j = 1, \ldots, k,$$

and hence,

$$|\lambda_j(\lambda_j - \lambda_*)| \le |\lambda_j|(|\lambda_j| + \varepsilon|\lambda_j|) \le (1+\varepsilon)\lambda_j^2.$$

Inserting this into (6.21) one obtains

$$\frac{2\nu}{\lambda_*^2} \ge \frac{1}{1+\varepsilon} \sum_{j=1}^{k} \frac{1}{\lambda_j^2} = \frac{2}{1+\varepsilon} |p''(0)| = \frac{\varepsilon^2}{1+\varepsilon} \frac{1}{\lambda_*^2}.$$

Thus, ε enjoys the quadratic inequality

$$\varepsilon^2 - 2\nu\varepsilon - 2\nu \le 0,$$

proving that $\varepsilon \le \nu + (\nu^2 + 2\nu)^{1/2}$. It therefore follows from (6.22) that

$$|\lambda_*| \le (\nu + (\nu^2 + 2\nu)^{1/2}) |2p''(0)|^{-1/2},$$

and hence,

$$|\lambda|^{2\nu} p(\lambda) \le |\lambda_*|^{2\nu} p(\lambda_*) \le c |p''(0)|^{-\nu}, \qquad \lambda_- \le \lambda \le \lambda_+,$$

because p is bounded by 1 in the given interval. $\qquad\square$

    This result enables the following proof of convergence of MR-II for the case of precise data.

Theorem 6.9 *If the right-hand side y belongs to $\mathcal{D}(T^\dagger)$ then the iterates x_k of MR-II converge to $T^\dagger y$ as $k \to \infty$.*

Proof. If $\kappa < \infty$ then the iteration terminates after $\kappa + 1$ steps and in this case $y - Tx_{\kappa+1} = p_{\kappa+1}(T)y = (I - P)y$ according to (6.7). Since $p'_{\kappa+1}(0) = q_\kappa(0) = 0$ it follows that $x_{\kappa+1} \perp \mathcal{N}(T)$, and hence, $x_{\kappa+1} = T^\dagger y$.
    Assume next that the iteration does not terminate, and assume further without loss of generality that $k \ge 2$. With $\lambda_{-,k}$ and $\lambda_{+,k}$ defined as before, $p_k/[(\lambda - \lambda_{-,k})(\lambda_{+,k} - \lambda)]$ is a polynomial in Π_{k-2}, and hence, by Proposition 6.1,

$$0 = [p_k, \frac{p_k}{(\lambda - \lambda_{-,k})(\lambda_{+,k} - \lambda)}]_2 = \int_{-\infty}^{\infty} p_k^2(\lambda) \frac{\lambda^2}{(\lambda - \lambda_{-,k})(\lambda_{+,k} - \lambda)} \, d\|E_\lambda y\|^2 .$$

Defining the function

$$\psi_k(\lambda) := \frac{\lambda^2}{(\lambda_{-,k} - \lambda)(\lambda_{+,k} - \lambda)} ,$$

and splitting the domain of integration into $\mathcal{I} = (\lambda_{-,k}, \lambda_{+,k}]$ and $\mathcal{I}_\infty = \mathbf{R} - \mathcal{I}$, the above can be rewritten as

$$\int_{\mathcal{I}_\infty} p_k^2(\lambda)\psi_k(\lambda) \, d\|E_\lambda y\|^2 = - \int_{\mathcal{I}} p_k^2(\lambda)\psi_k(\lambda) \, d\|E_\lambda y\|^2 . \tag{6.23}$$

Consider next the function ψ_k. Obviously, $\psi_k(\lambda)$ is negative for $\lambda \in \mathcal{I}$, and positive for $\lambda \in \mathcal{I}_\infty$. If $\lambda_{-,k} + \lambda_{+,k} = 0$, i.e., if ψ_k is even then $\psi_k(\lambda) > 1$ for all $\lambda \in \mathcal{I}_\infty$. If $\lambda_{-,k} + \lambda_{+,k} \neq 0$ then the situation is more complicated: although $\psi_k(\lambda)$ tends to one as $|\lambda| \to \infty$, this is no bound for ψ_k in $\mathcal{I}_\infty$. Instead, elementary calculus shows that ψ_k has precisely one local extremum in $\mathcal{I}_\infty$, namely a local minimum at

$$\lambda_{0,k} := 2 \frac{\lambda_{+,k}\lambda_{-,k}}{\lambda_{-,k} + \lambda_{+,k}} = 2(\frac{1}{\lambda_{-,k}} + \frac{1}{\lambda_{+,k}})^{-1} . \tag{6.24}$$

By formally setting $\lambda_{0,k} := \infty$ and $\psi_k(\infty) := 1$ in the case when $\lambda_{-,k} + \lambda_{+,k} = 0$, one therefore always has

$$\psi_k(\lambda) \geq \psi_k(\lambda_{0,k}) > 0, \qquad \lambda \in \mathcal{I}_\infty .$$

Using this information about ψ_k, the integrand on the left-hand side of (6.23) can be estimated from below by $p_k^2\psi_k(\lambda_{0,k})$, which gives

$$\int_{\mathcal{I}_\infty} p_k^2(\lambda) \, d\|E_\lambda y\|^2 \leq - \int_{\mathcal{I}} p_k^2(\lambda) \frac{\psi_k(\lambda)}{\psi_k(\lambda_{0,k})} \, d\|E_\lambda y\|^2 .$$

Consequently,

$$\|y - Tx_k\|^2 = [p_k, p_k]_0 = \int_{\mathcal{I}} p_k^2(\lambda) \, d\|E_\lambda y\|^2 + \int_{\mathcal{I}_\infty} p_k^2(\lambda) \, d\|E_\lambda y\|^2$$

$$\leq \int_{\mathcal{I}} p_k^2(\lambda)\,(1 - \frac{\psi_k(\lambda)}{\psi_k(\lambda_{0,k})}) \, d\|E_\lambda y\|^2 .$$

Since $\psi_k(\lambda)$ is negative for $\lambda \in \mathcal{I}$ and $\psi_k(\lambda_{0,k}) > 0$, the nonnegative function

$$\varphi_k(\lambda) := p_k(\lambda) \left(1 - \frac{\psi_k(\lambda)}{\psi_k(\lambda_{0,k})}\right)^{1/2}, \qquad \lambda \in \mathcal{I}. \tag{6.25}$$

is well-defined. Introducing the spectral orthoprojector

$$E_\mathcal{I} := E_{\lambda_{+,k}} - E_{\lambda_{-,k}} , \tag{6.26}$$

103

which corresponds to spectral elements $\lambda \in \mathcal{I}$, the above inequality can be rewritten as

$$\|y - Tx_k\| \le \|E_{\mathcal{I}}\varphi_k(T)y\| .\qquad (6.27)$$

Up to this point the assumption $y \in \mathcal{D}(T^\dagger)$ has not been used. Now let $x = T^\dagger y$, so that $Tx = Py$. The two quantities $y - Tx_k = p_k(T)y$ and $E_{\mathcal{I}}\varphi_k(T)y$ in (6.27) both have the component $(I - P)y$ in $\mathcal{N}(T)$, and can only differ by their component in $\mathcal{R}(T) = \mathcal{N}(T)^\perp$. By Pythagoras' theorem inequality (6.27) is therefore equivalent to

$$\|Tx - Tx_k\| \le \|E_{\mathcal{I}}\varphi_k(T)Tx\| .\qquad (6.28)$$

The function φ_k^2 of (6.25) needs a more detailed study. Obviously, $\varphi_k^2(0) = 1$, and the poles of ψ_k are cancelled by zeros of p_k^2. Consequently, φ_k^2 can be extended to a polynomial in Π_{2k}^0. To see that φ_k^2 actually belongs to Π_{2k}^{00}, consider its derivative

$$(\varphi_k^2)' = 2p_k p_k'(1 - \frac{\psi_k}{\psi_k(\lambda_{0,k})}) - p_k^2 \frac{\psi_k'}{\psi_k(\lambda_{0,k})} .$$

Since $p_k'(0) = \psi_k'(0) = 0$, this establishes $(\varphi_k^2)'(0) = 0$, and hence, $\varphi_k^2 \in \Pi_{2k}^{00}$. As can be seen from (6.25), $\lambda_{-,k}$ and $\lambda_{+,k}$ are simple zeros of φ_k^2; every further zero of p_k is a double zero of φ_k^2; finally, the remaining two zeros of φ_k^2 are the roots of the equation $\psi_k(\lambda) = \psi_k(\lambda_{0,k})$ counting multiplicity. As $\lambda_{0,k}$ is a local minimum of ψ_k, one has $\psi_k'(\lambda_{0,k}) = 0$, and hence, $\lambda_{0,k}$ is the missing double root of φ_k^2. These are all $2k$ zeros of φ_k^2. In the special case $\lambda_{-,k} + \lambda_{+,k} = 0$, φ_k^2 degenerates to a polynomial of degree $2k - 2$, and all zeros of φ_k^2 are zeros of p_k.

The above discussion shows that φ_k^2 satisfies the assumptions of Lemma 6.8. Consequently, for any $\nu > 0$ there is some $c > 0$ with

$$|\lambda|^{2\nu}\varphi_k^2(\lambda) \le c |(\varphi_k^2)''(0)|^{-\nu} \le c |p_k''(0)|^{-\nu}, \qquad \lambda \in \mathcal{I},\qquad (6.29)$$

where for the second inequality in (6.29) it has been used that $|(\varphi_k^2)''(0)| \ge |p_k''(0)|$. This follows readily from (6.18) and the above identification of all zeros of φ_k^2:

$$|(\varphi_k^2)''(0)| = \frac{1}{2}(2 \sum_{j=1}^{k} \frac{1}{\lambda_{j,k}^2} - \frac{1}{\lambda_{-,k}^2} - \frac{1}{\lambda_{+,k}^2} + 2\frac{1}{\lambda_{0,k}^2}) \ge \frac{1}{2} \sum_{j=1}^{k} \frac{1}{\lambda_{j,k}^2} = |p_k''(0)| .$$

To complete the proof, assume without loss of generality that $|\lambda_{-,k}| \le \lambda_{+,k}$, and choose ε (depending on k) so as to fulfill $\varepsilon \le |\lambda_{-,k}|$. Denote by $P_\varepsilon := E_\varepsilon - E_{-\varepsilon}$ the spectral orthoprojector corresponding to the interval $(-\varepsilon, \varepsilon]$. Recall that $p_k \in \Pi_k^{00}$, which implies that the associated iteration polynomial q_{k-1} satisfies $q_{k-1}(0) = 0$, and hence, with $x = T^\dagger y$,

$$x - x_k = x - q_{k-1}(T)y = x - q_{k-1}(T)Tx - q_{k-1}(0)(I - P)y = p_k(T)x .$$

Therefore, the iteration error can be estimated as

104

$$\begin{aligned} \|x - x_k\| &\leq \|P_\varepsilon(x - x_k)\| + \|(I - P_\varepsilon)(x - x_k)\| \\ &\leq \|P_\varepsilon p_k(T)x\| + \frac{1}{\varepsilon}\|(I - P_\varepsilon)T(x - x_k)\| . \end{aligned}$$

Since all roots of p_k are real, and since $p_k \in \Pi_k^{00}$, p_k is bounded by one in $(-\varepsilon, \varepsilon] \subset \mathcal{I}$. Together with (6.28) this yields

$$\|x - x_k\| \leq \|P_\varepsilon x\| + \frac{1}{\varepsilon}\|T(x - x_k)\| \leq \|P_\varepsilon x\| + \frac{1}{\varepsilon}\|E_\mathcal{I}\varphi_k(T)Tx\| . \qquad (6.30)$$

Still, there is freedom in choosing ε in an optimal way, subject to $\varepsilon \leq |\lambda_{-,k}|$. If $\lambda_{-,k}$ and $\lambda_{+,k}$ both tend to zero as $k \to \infty$, then let $\varepsilon := |\lambda_{-,k}|$. Inserting (6.29) with $\nu = 1$ into (6.30) then gives

$$\|x - x_k\| \leq \|P_\varepsilon x\| + c\frac{|p_k''(0)|^{-1/2}}{\varepsilon}\|E_\mathcal{I}x\| \leq (1 + \sqrt{2}c)\|E_\mathcal{I}x\| \longrightarrow 0, \qquad k \to \infty,$$

since $|p_k''(0)|^{-1/2} \leq \sqrt{2}\,|\lambda_{-,k}| = \sqrt{2}\,\varepsilon$ by virtue of (6.18). If neither $\lambda_{-,k}$ nor $\lambda_{+,k}$ tend to zero, one may choose $\varepsilon := |p_k''(0)|^{-1/4}$. Note that this implies that ε goes to zero as $k \to \infty$ by virtue of (6.19). With this choice of ε and with k sufficiently large, (6.30) and (6.29) with $\nu = 1$ give

$$\|x - x_k\| \leq \|P_\varepsilon x\| + c\frac{|p_k''(0)|^{-1/2}}{\varepsilon}\|E_\mathcal{I}x\| = \|P_\varepsilon x\| + c\varepsilon\|E_\mathcal{I}x\| \longrightarrow 0, \qquad k \to \infty.$$

The final case in which $\lambda_{-,k} \to 0$ as $k \to \infty$, but $\lambda_{+,k}$ won't (recall that $|\lambda_{-,k}| \leq \lambda_{+,k}$ by assumption) is somewhat more difficult to handle. In this case let $\varepsilon := |\lambda_{-,k}|$ and observe that $\sqrt{\varepsilon}$ will be smaller than $\lambda_{+,k}$ for k sufficiently large. The interval $\mathcal{I}$ can then be split into two intervals $\mathcal{I}_1 = (-\varepsilon, \sqrt{\varepsilon}]$ and $\mathcal{I}_2 = (\sqrt{\varepsilon}, \lambda_{+,k}]$; denote by $E_{\mathcal{I}_1}$ and $E_{\mathcal{I}_2}$ the corresponding two spectral orthoprojectors. Splitting the right-hand side of (6.30) accordingly, one obtains

$$\begin{aligned} \|x - x_k\| &\leq \|P_\varepsilon x\| + \frac{1}{\varepsilon}\|E_{\mathcal{I}_1}\varphi_k(T)Tx\| + \frac{1}{\varepsilon}\|E_{\mathcal{I}_2}\varphi_k(T)Tx\| \\ &\leq \|P_\varepsilon x\| + \frac{1}{\varepsilon}\|E_{\mathcal{I}_1}\varphi_k(T)Tx\| + \frac{1}{\varepsilon^{3/2}}\|E_{\mathcal{I}_2}\varphi_k(T)T^2x\| . \end{aligned}$$

Now (6.29) can be applied with $\nu = 1$ and $\nu = 2$, respectively. This gives

$$\begin{aligned} \|x - x_k\| &\leq \|P_\varepsilon x\| + c\frac{|p_k''(0)|^{-1/2}}{\varepsilon}\|E_{\mathcal{I}_1}x\| + c\frac{|p_k''(0)|^{-1}}{\varepsilon^{3/2}}\|E_{\mathcal{I}_2}x\| \\ &\leq \|P_\varepsilon x\| + \sqrt{2}c\,\|P_{\sqrt{\varepsilon}}x\| + 2c\,\sqrt{\varepsilon}\,\|x\| , \end{aligned}$$

and all three terms on the right-hand side go to zero as $k \to \infty$. Thus, convergence $x_k \to x = T^\dagger y$ has been established and the proof is complete. $\qquad \square$

It will be shown next that the MR-II iterates diverge to infinity in norm when the assumption of Theorem 6.9 is not fulfilled.

Theorem 6.10 *Let $\{x_k\}$ be the iterates of* MR-II. *If $y \notin \mathcal{D}(T^\dagger)$ then $\|x_k\| \to \infty$ as $k \to \infty$.*

Proof. The proof is very similar to the proof of Theorem 3.5. If $y \notin \mathcal{D}(T^\dagger)$ then $\kappa = \infty$ and hence the iteration does not terminate. For $k \geq 1$ define $\varphi_{2k}(\lambda) = (1 - \lambda^2)^k \in \Pi_{2k}^{00}$ and let $\varphi_{2k+1} = \varphi_{2k}$. The sequence $\{\varphi_k\}$ is uniformly bounded, and converges pointwise to zero on $[-1, 1] \setminus \{0\}$. Consequently, $\varphi_k(T)y$ converges to $(I - P)y$ as $k \to \infty$. By definition of the residual polynomials $\{p_k\}$, this implies that

$$\limsup_{k \to \infty} \|p_k(T)y\| \leq \lim_{k \to \infty} \|\varphi_k(T)y\| = \|(I - P)y\| .$$

Since

$$\|p_k(T)y\|^2 = \|y - Tx_k\|^2 = \|(I - P)y\|^2 + \|Py - Tx_k\|^2 ,$$

the above is equivalent to

$$\|Py - Tx_k\| \to 0, \qquad k \to \infty . \tag{6.31}$$

Assume now that some subsequence of $\{x_k\}$ remains bounded, so that it has a weakly converging subsequence with weak limit x, say. It follows that the images of this subsequence converge weakly to Tx and, because of (6.31), this yields $Tx = Py$. Consequently, $y \in \mathcal{D}(T^\dagger)$, and hence, $\|x_k\|$ must go to infinity if $y \notin \mathcal{D}(T^\dagger)$. ☐

6.4 Stopping rules for MR-II

The minimization property of MR-II suggests that the discrepancy principle (i.e., Stopping Rule 3.10) is the correct stopping rule for this algorithm. This is indeed the case. A number of technical difficulties arise, though, when generalizing the proofs from Section 3.3. One of them is the estimation of the maximum modulus of the iteration polynomial q_{k-1} in the vicinity of the origin. In the context of (semi)definite problems, cf. Section 3.3, the corresponding estimate followed comparatively easy from the convexity of the residual polynomial and its derivative, cf. (3.14). Here the analysis is more difficult and will be the subject of the next lemma. Recall that $\lambda_{-,k}$ and $\lambda_{+,k}$ are the largest negative and the smallest positive root of p_k, respectively. As in the previous section let $\mathcal{I} = (\lambda_{-,k}, \lambda_{+,k}]$ and $E_\mathcal{I}$ be the corresponding spectral orthoprojector, cf. (6.26).

Lemma 6.11 *For all $k \in \mathbf{N}$ the following inequality holds:*

$$|q_{k-1}(\lambda)| \leq 2 |p_k''(0)|^{1/2}, \qquad \lambda_{-,k} \leq \lambda \leq \lambda_{+,k} . \tag{6.32}$$

Proof. For $k = 0$ and $k = 1$ the lemma is clearly true since the iteration polynomial vanishes identically. Next fix $k \geq 2$, and denote by $\lambda = \lambda_* \in \bar{\mathcal{I}}$ one of the points where $|q_{k-1}(\lambda)|$ attains its maximum. Here, $\bar{\mathcal{I}}$ denotes the closure of $\mathcal{I}$. Clearly, $\lambda_* \neq 0$ since $q_{k-1}(0) = 0$. Furthermore, let

$$\varepsilon := |2p_k''(0)|^{1/2} / |q_{k-1}(\lambda_*)| . \tag{6.33}$$

If $\varepsilon \geq 1$ for all local maxima $\lambda_* \in \bar{\mathcal{I}}$ then (6.32) holds true. Assume therefore that $\varepsilon < 1$ for some $\lambda_* \in \bar{\mathcal{I}}$. Inserting the definition of q_{k-1} and using that $0 \leq p_k(\lambda) \leq 1$ for $\lambda \in \bar{\mathcal{I}}$, it follows from (6.33) that

$$\varepsilon = |2p_k''(0)|^{1/2} \frac{|\lambda_*|}{1 - p_k(\lambda_*)} \geq |2p_k''(0)|^{1/2} |\lambda_*| .$$

Consequently, cf. (6.18),

$$|\lambda_*| \leq \varepsilon |2p_k''(0)|^{-1/2} \leq \varepsilon |\lambda_{j,k}| \qquad \text{for every } j = 1, \ldots, k . \tag{6.34}$$

Since $\varepsilon < 1$ this shows that λ_* is an interior point of the interval $\mathcal{I}$; being a local extremum of q_{k-1} by construction this implies further that $q_{k-1}'(\lambda_*) = 0$. Straightforward calculus yields

$$q_{k-1}' = -p_k \left(\frac{q_{k-1}}{\lambda p_k} + \frac{p_k'}{\lambda p_k} \right) ,$$

and hence,

$$\left| \frac{p_k'(\lambda_*)}{\lambda_* p_k(\lambda_*)} \right| = \left| \frac{q_{k-1}(\lambda_*)}{\lambda_* p_k(\lambda_*)} \right| .$$

Substituting $\lambda_* = (1 - p_k(\lambda_*))/q_{k-1}(\lambda_*)$ on the right-hand side and using (6.33) one obtains

$$\left| \frac{p_k'(\lambda_*)}{\lambda_* p_k(\lambda_*)} \right| = \left| \frac{q_{k-1}^2(\lambda_*)}{p_k(\lambda_*)(1 - p_k(\lambda_*))} \right| = 2 \frac{|p_k''(0)|}{\varepsilon^2} \frac{1}{p_k(\lambda_*)(1 - p_k(\lambda_*))} ,$$

and, since $0 \leq t(1 - t) \leq 1/4$ for $0 \leq t \leq 1$, this yields

$$\left| \frac{p_k'(\lambda_*)}{\lambda_* p_k(\lambda_*)} \right| \geq \frac{8}{\varepsilon^2} |p_k''(0)| . \tag{6.35}$$

Note that, in view of (6.13),

$$\frac{p_k'(\lambda)}{\lambda p_k(\lambda)} = \sum_{j=1}^{k} \frac{1}{\lambda(\lambda - \lambda_{j,k})} = \sum_{j=1}^{k} \frac{1}{\lambda_{j,k}} \left(\frac{1}{\lambda - \lambda_{j,k}} - \frac{1}{\lambda} \right) = \sum_{j=1}^{k} \frac{1}{\lambda_{j,k}} \frac{1}{\lambda - \lambda_{j,k}} .$$

Using (6.34), the inverse triangle inequality gives

$$|\lambda_{j,k}(\lambda_* - \lambda_{j,k})| \geq |\lambda_{j,k}|(|\lambda_{j,k}| - \varepsilon |\lambda_{j,k}|) = (1 - \varepsilon)\lambda_{j,k}^2 ,$$

107

and hence,

$$\left| \frac{p_k'(\lambda_*)}{\lambda_* p_k(\lambda_*)} \right| \le \frac{1}{1-\varepsilon} \sum_{j=1}^{k} \frac{1}{\lambda_{j,k}^2} = \frac{2}{1-\varepsilon} |p_k''(0)| .$$

Combining this with (6.35) one obtains $8\varepsilon^{-2} \le 2(1-\varepsilon)^{-1}$, which gives $\varepsilon \ge 2(\sqrt{2}-1)$. From (6.33) therefore follows

$$|q_{k-1}(\lambda)| \le \frac{\sqrt{2}}{2(\sqrt{2}-1)} |p_k''(0)|^{1/2} = \frac{1}{2} (2+\sqrt{2}) |p_k''(0)|^{1/2} ,$$

which is somewhat stronger than (6.32).

The remaining two preparatory lemmas are organized in much the same way as Section 3.2. First, in Lemma 6.12, an estimate for the rate with which the residuals decrease will be derived on the basis of the convergence analysis in Theorem 6.9. Afterwards, in Lemma 6.13, an inequality for the iteration error in terms of the residual norm and the modulus of $p_k''(0)$ is given. The proof of the order-optimality of the discrepancy principle in Theorem 6.15 amounts to an estimation of this second derivative.

Lemma 6.12 *Consider the iterates* $\{x_k^\delta\}$ *of* MR-II *with respect to some perturbed right-hand side* y^δ. *If* $y \in \mathcal{R}(T)$ *and* $\kappa = \infty$ *then*

$$\limsup_{k \to \infty} \|y^\delta - Tx_k^\delta\| \le \|y - y^\delta\| .$$

If $y \in \mathcal{R}(T)$ *satisfies Assumption 3.6 then, for* $2 \le k \le \kappa + 1$,

$$\|y^\delta - Tx_k^\delta\| \le \|y - y^\delta\| + c |p_k''(0)|^{-(\mu+1)/2} \omega .$$

Proof. In the proof of Theorem 6.9 it has been shown, cf. (6.27), that

$$\|y^\delta - Tx_k^\delta\| \le \|E_\mathcal{I}\varphi_k(T)y^\delta\| ,$$

with φ_k as in (6.25) and $E_\mathcal{I} = E_{\lambda_{+,k}} - E_{\lambda_{-,k}}$. In fact, this result has been established there not only for $y^\delta \in \mathcal{D}(T^\dagger)$ but for any right-hand side $y^\delta \in \mathcal{X}$. Furthermore, φ_k^2 has been shown to be a polynomial in Π_{2k}^{00}, with no zero in $\mathcal{I} = (\lambda_{-,k}, \lambda_{+,k}]$. It follows that $0 \le \varphi_k(\lambda) \le 1$ for $\lambda \in \mathcal{I}$, and hence,

$$\|y^\delta - Tx_k^\delta\| \le \|E_\mathcal{I}\varphi_k(T)(y^\delta - y)\| + \|E_\mathcal{I}\varphi_k(T)y\| \le \|y^\delta - y\| + \|E_\mathcal{I}\varphi_k(T)y\| .$$

Since $y = Tx$ for some $x \in \mathcal{X}$, (6.29) with $\nu = 1$ yields

$$\|E_\mathcal{I}\varphi_k(T)y\| = \|E_\mathcal{I}\varphi_k(T)Tx\| \le c |p_k''(0)|^{-1/2} \|E_\mathcal{I}x\| \le c |p_k''(0)|^{-1/2} \|x\| ,$$

where the right-hand side goes to zero as $k \to \infty$, cf. (6.19). This proves the first claim.

If Assumption 3.6 is satisfied then $y = T^{\mu+1}w$ for some $w \in \mathcal{X}$ with $\|w\| = \omega$, and one can apply (6.29) with $\nu = \mu + 1$. This yields

$$\|E_\mathcal{I}\varphi_k(T)y\| = \|E_\mathcal{I}\varphi_k(T)T^{\mu+1}w\| \leq c\,|p_k''(0)|^{-(\mu+1)/2}\,\|E_\mathcal{I}w\|\,,$$

from which the second assertion follows. □

Lemma 6.13 *Let y satisfy Assumption 3.6. Then the iteration error of MR-II applied to the perturbed right-hand side y^δ satisfies*

$$\|T^\dagger y - x_k^\delta\| \leq c\,(\omega^{1/\mu+1}\rho_k^{\mu/\mu+1} + |p_k''(0)|^{1/2}\,\|y - y^\delta\|) \tag{6.36}$$

for $0 \leq k \leq \kappa + 1$, with

$$\rho_k := \max\{\|y^\delta - Tx_k^\delta\|, \|y - y^\delta\|\}\,. \tag{6.37}$$

Proof. For $k = 0$ and $k = 1$ one has $x_k^\delta = 0$ and (6.36) follows as in the proof of Lemma 3.8. For $k \geq 2$ let ε be such that

$$0 < \varepsilon \leq |2p_k''(0)|^{-1/2}\,.$$

By virtue of (6.18) this implies that $\varepsilon \leq |\lambda_{j,k}|$, $j = 1, \dots, k$. Let

$$P_\varepsilon = E_\varepsilon - E_{-\varepsilon}$$

be the spectral orthoprojector for the interval $(-\varepsilon, \varepsilon]$. As in the proof of Lemma 3.8 let

$$\tilde{x}_k = q_{k-1}(T)y\,,$$

so that $x - \tilde{x}_k = p_k(T)x$ for $x = T^\dagger y$. This gives rise to the following estimate:

$$\begin{aligned}
\|x - x_k^\delta\| &\leq \|P_\varepsilon(x - x_k^\delta)\| + \|(I - P_\varepsilon)(x - x_k^\delta)\| \\
&\leq \|P_\varepsilon(x - \tilde{x}_k)\| + \|P_\varepsilon(\tilde{x}_k - x_k^\delta)\| + \varepsilon^{-1}\|(I - P_\varepsilon)(y - Tx_k^\delta)\| \\
&\leq \|P_\varepsilon p_k(T)T^\mu w\| + \|P_\varepsilon q_{k-1}(T)(y - y^\delta)\| + \varepsilon^{-1}\|y - Tx_k^\delta\| \\
&\leq \|\lambda^\mu p_k(\lambda)\|_{[-\varepsilon,\varepsilon]}\,\omega + \|q_{k-1}(\lambda)\|_{[-\varepsilon,\varepsilon]}\,\|y - y^\delta\| \\
&\quad + \varepsilon^{-1}(\|y^\delta - Tx_k^\delta\| + \|y - y^\delta\|)\,.
\end{aligned}$$

By construction of ε, p_k is bounded by one in $[-\varepsilon, \varepsilon]$, and Lemma 6.11 provides a bound for q_{k-1} in the same interval. Using the definition (6.37) of ρ_k, this gives

$$\|x - x_k^\delta\| \leq \varepsilon^\mu \omega + 2\varepsilon^{-1}\rho_k + 2|p_k''(0)|^{1/2}\,\|y - y^\delta\|\,. \tag{6.38}$$

Note that (6.38) coincides with (3.15) from the proof of Lemma 3.8, when replacing $|p'_k(0)|$ in (3.15) by $2|p''_k(0)|^{1/2}$. With this substitution in mind the remainder of the proof of Lemma 3.8 can be copied almost word by word to obtain the desired conclusion (6.36). Of course, ρ_k must be estimated here by Lemma 6.12 instead of Lemma 3.7. $\square$

Note the pronounced role of $|p''_k(0)|^{1/2}$ in the bounds for residual norm and iteration error. However, unlike the sequence $\{|p'_k(0)|\}$ whose elements had a similar role in the previous chapters, the sequence $\{|p''_k(0)|^{1/2}\}$ need not increase monotonically:

Example 6.14 Let $y = \varepsilon v_1 + v_2$, where v_1 and v_2 are eigenvectors of T corresponding to eigenvalues $\lambda_1 = 1$ and $\lambda_2 = -1/2$, respectively. In this case one has $\kappa = 2$, and $[p_3, p_3]_0 = 0$ for
$$p_3(\lambda) = (1 - \lambda)^2(1 + 2\lambda) \in \Pi_3^{00}.$$
Note that $|p''_3(0)| = 6$. The polynomial $p_2 \in \Pi_2^{00}$ minimizing $[p_2, p_2]_0$ must have the form $p_2(\lambda) = 1 - (\lambda/\lambda_{-,2})^2$, and it is obvious that its negative zero $\lambda_{-,2}$ approaches $\lambda_2 = -1/2$ as $\varepsilon \to 0$. Hence, $|p''_2(0)| \approx 8$ for ε close to zero.

As will be shown next, this slightly different situation does not affect the order-optimality of the discrepancy principle.

Theorem 6.15 *Let $y \in \mathcal{R}(T)$ and assume that $\|y - y^\delta\| \leq \delta$. Then Stopping Rule 3.10 determines a finite stopping index $k(\delta)$ for* MR-II *with*
$$x^\delta_{k(\delta)} \to T^\dagger y, \qquad \delta \to 0. \tag{6.39}$$

If in addition y satisfies Assumption 3.6 then
$$\|T^\dagger y - x^\delta_{k(\delta)}\| \leq c\omega^{1/\mu+1}\delta^{\mu/\mu+1}. \tag{6.40}$$

Proof. If $\kappa = \infty$ then the existence of a finite stopping index $k(\delta)$ follows from Lemma 6.12. If κ is finite then there is also a well-defined stopping index $k(\delta) \leq \kappa+1$, since $[p_{\kappa+1}, p_{\kappa+1}]_0 = \|(I - P)y^\delta\|^2 \leq \delta^2$ for $y \in \mathcal{R}(T)$. In the sequel, the argument δ in $k(\delta)$ will be omitted for the ease of notation.

Assume first that y satisfies Assumption 3.6. Since $k \leq \kappa + 1$, Lemma 6.13 applies and yields the following bound for the error at the stopping index:
$$\|T^\dagger y - x^\delta_k\| \leq c(\omega^{1/\mu+1}\delta^{\mu/\mu+1} + |p''_k(0)|^{1/2}\delta). \tag{6.41}$$

If $k \leq 1$ then $p''_k(0) = 0$, which implies (6.40) for this case. It remains to estimate $p''_k(0)$ when $k \geq 2$. When $k > 2$ Lemma 6.12 and Stopping Rule 3.10 yield
$$\tau\delta < \|y^\delta - Tx^\delta_{k-1}\| \leq \delta + c|p''_{k-1}(0)|^{-(\mu+1)/2}\omega,$$

110

and hence,

$$|p''_{k-1}(0)| \leq c\,(\omega/\delta)^{2/\mu+1}\,. \tag{6.42}$$

Note that this is trivially fulfilled when $k = 2$. (6.42) and (6.41) already imply (6.40) in the case when $u^{[4]}_{k-2}(0) = 0$ since then $p''_k(0) = p''_{k-1}(0)$ by virtue of Lemma 6.4. Therefore assume in the sequel that $u^{[4]}_{k-2}(0) \neq 0$, in which case $p^{[4]}_{k-2} = u^{[4]}_{k-2}/u^{[4]}_{k-2}(0)$ exists. In this final case the proof of (6.40) is much harder and shall be done in three steps.

Step 1. The aim of the first step is to prove the existence of some constant $c > 0$ with

$$|p''_k(0)|^2 \leq c\left((\omega/\delta)^{4/\mu+1} + \frac{[p_{k-1},p_{k-1}]_0}{[p^{[4]}_{k-2},p^{[4]}_{k-2}]_4}\right). \tag{6.43}$$

By assumption the right-hand side of (6.43) is well-defined, and

$$[p^{[4]}_{k-2},p^{[4]}_{k-2}]_4 = |u^{[4]}_{k-2}(0)|^{-2}[u^{[4]}_{k-2},u^{[4]}_{k-2}]_4 = |u^{[4]}_{k-2}(0)|^{-2}\,.$$

Therefore Lemma 6.4 yields the representation

$$|p''_k(0) - p''_{k-1}(0)|^2 = 4\,\frac{[p_{k-1},p_{k-1}]_0 - [p_k,p_k]_0}{[p^{[4]}_{k-2},p^{[4]}_{k-2}]_4} \leq 4\,\frac{[p_{k-1},p_{k-1}]_0}{[p^{[4]}_{k-2},p^{[4]}_{k-2}]_4}\,,$$

and (6.43) follows from (6.42) and the standard inequality

$$|p''_k(0)|^2 \leq 2(|p''_{k-1}(0)|^2 + |p''_k(0) - p''_{k-1}(0)|^2)\,.$$

Step 2. Let τ be the parameter occurring in Stopping Rule 3.10, and assume for the moment that there are constants c_1 and c_2 with

$$0 < c_1 < c_2 < \tau - 1\,, \tag{6.44}$$

such that the following holds: for some ε with

$$(c_1\,\delta/\omega)^{1/\mu+1} \leq \varepsilon \leq (c_2\,\delta/\omega)^{1/\mu+1}\,, \tag{6.45}$$

there is a polynomial $\psi \in \Pi^{00}_{k-1}$ with the following two properties:

$$\psi \text{ has no zeros in } [-\varepsilon,\varepsilon]\,,$$
$$|\psi(\lambda)| \leq c\varepsilon^{-2}|\lambda^2 p^{[4]}_{k-2}(\lambda)|, \qquad |\lambda| \geq \varepsilon\,. \tag{6.46}$$

The purpose of this second step is to show that

$$[p_{k-1},p_{k-1}]_0 \leq c(\omega/\delta)^{4/\mu+1}[p^{[4]}_{k-2},p^{[4]}_{k-2}]_4\,, \tag{6.47}$$

provided conditions (6.44) - (6.46) hold. These conditions will be established in the third step.

The proof of (6.47) is similar to the second step in the proof of Theorem 3.11. From the optimality property of p_{k-1} follows that

$$[p_{k-1}, p_{k-1}]_0 \le [\psi, \psi]_0 = \|P_\varepsilon \psi(T) y^\delta\|^2 + \|(I - P_\varepsilon) \psi(T) y^\delta\|^2, \qquad (6.48)$$

where, once again, P_ε denotes the spectral projector associated with the interval $(-\varepsilon, \varepsilon]$. As $\psi \in \Pi^{00}_{k-1}$, and since ψ has no zeros in $[-\varepsilon, \varepsilon]$, $0 \le \psi(\lambda) \le 1$ holds true for $\lambda \in [-\varepsilon, \varepsilon]$, and hence, using Assumption 3.6 and the second inequality in (6.45), the first term on the right-hand side can be estimated by

$$\|P_\varepsilon \psi(T) y^\delta\| \le \|P_\varepsilon y^\delta\| \le \|P_\varepsilon T^{\mu+1} w\| + \|P_\varepsilon (y^\delta - y)\| \le \varepsilon^{\mu+1} \omega + \delta \le (c_2 + 1)\delta.$$

The second term on the right-hand side of (6.48) can be estimated with the second condition in (6.46):

$$\|(I - P_\varepsilon) \psi(T) y^\delta\|^2 \le \frac{c^2}{\varepsilon^4} \|(I - P_\varepsilon) T^2 p^{[4]}_{k-2}(T) y^\delta\|^2 \le \frac{c^2}{\varepsilon^4} [p^{[4]}_{k-2}, p^{[4]}_{k-2}]_4.$$

Inserting these two results into (6.48), one obtains

$$\begin{aligned} [p_{k-1}, p_{k-1}]_0 &\le \frac{(c_2 + 1)^2}{\tau^2} (\tau\delta)^2 + \frac{c^2}{\varepsilon^4} [p^{[4]}_{k-2}, p^{[4]}_{k-2}]_4 \\ &\le \frac{(c_2 + 1)^2}{\tau^2} [p_{k-1}, p_{k-1}]_0 + \frac{c^2}{\varepsilon^4} [p^{[4]}_{k-2}, p^{[4]}_{k-2}]_4, \end{aligned}$$

where the latter follows from the definition of the stopping index $k = k(\delta)$. Note that $(c_2 + 1)/\tau < 1$ by assumption, cf. (6.44); hence, gathering terms one obtains a new constant $c > 0$ for which

$$[p_{k-1}, p_{k-1}]_0 \le c \varepsilon^{-4} [p^{[4]}_{k-2}, p^{[4]}_{k-2}]_4.$$

(6.47) now follows from the left-hand inequality for ε in (6.45).

Step 3. In this step c_1, c_2, ε, and the corresponding polynomial ψ are constructed for which (6.44) – (6.46) hold. Since $|\lambda_{j,k-1}| \ge (2|p''_{k-1}(0)|)^{-1/2}$ for all $1 \le j \le k - 1$, it follows from (6.42) that there exists $0 < c_2 < \tau - 1$ with

$$|\lambda_{j,k-1}| \ge 4 (c_2 \delta/\omega)^{1/\mu+1}, \qquad j = 1, \ldots, k - 1. \qquad (6.49)$$

Recall from Section 6.2 (namely from the discussion preceeding Lemma 6.7) that all zeros $\mu_{j,k-2}$ of $p^{[4]}_{k-2}$ ($u^{[4]}_{k-2}$, respectively) in $[\lambda_{-,k}, \lambda_{+,k}]$ belong to the exceptional set $\mathcal{E}_{k-2}$, and $\mathcal{E}_{k-2}$ has at most two elements. The location of the elements of $\mathcal{E}_{k-2}$ determines the actual value of ε for which a polynomial ψ satisfying (6.46) will be constructed. ε is determined in such a way that the following condition holds true:

$$\{\lambda \mid \varepsilon/2 < |\lambda| \le 4\varepsilon\} \cap \mathcal{E}_{k-2} = \emptyset. \qquad (6.50)$$

Choosing $\varepsilon = (c_2\delta/\omega)^{1/\mu+1}$ the left-hand side of (6.50) is contained in $[\lambda_{-,k}, \lambda_{+,k}]$ and hence can contain at most two elements by virtue of (6.49). Therefore, if (6.50) is not already satisfied for $\varepsilon = (c_2\delta/\omega)^{1/\mu+1}$ then ε is subsequently replaced by $\varepsilon/8$ and $\varepsilon/64$, respectively, until the criterion (6.50) is finally met. Not more than two such correction steps are necessary to determine a value of ε for which (6.50) actually holds true, and this ε satisfies (6.45) with $c_1 = c_2/64$. Note that c_1 and c_2 only depend on τ and on the constant c in (6.42).

The polynomial ψ is constructed from the following Ansatz:

$$\psi(\lambda) = \frac{(1-\gamma\lambda)p_{k-2}^{[4]}(\lambda)}{\prod_{|\mu_{j,k-2}|\leq 4\varepsilon}(1-\lambda/\mu_{j,k-2})}. \tag{6.51}$$

Recall that $\mu_{j,k-2}$, $j = 1, \ldots, k-2$, are the zeros of $p_{k-2}^{[4]}$. Consequently ψ is a polynomial of degree $k-1$ or less, and $\psi \in \Pi_{k-1}^{00}$ holds true, if and only if

$$\gamma = p_{k-2}^{[4]\prime}(0) + \sum_{|\mu_{j,k-2}|\leq 4\varepsilon} \frac{1}{\mu_{j,k-2}} = -\sum_{|\mu_{j,k-2}|>4\varepsilon} \frac{1}{\mu_{j,k-2}}.$$

The zeros $\mu_{j,k-2}$ occuring on the right-hand side include all $\mu_{j,k-2} \notin \mathcal{E}_{k-2}$, together with those belonging to $\mathcal{E}_{k-2} \setminus [-4\varepsilon, 4\varepsilon]$. Since $\mathcal{E}_{k-2}$ has not more than two elements, the contribution to the right-hand side coming from those latter zeros is at most two times $1/(4\varepsilon)$ in absolute value, whereas a bound for the contribution coming from the remaining zeros $\mu_{j,k-2} \notin \mathcal{E}_{k-2}$ has been obtained in Lemma 6.7, cf. (6.15). Note that $k \leq \kappa + 1$, and hence $k-2 < \kappa$ as required. In terms of the notation (6.14) of Section 6.2 this yields the following bound for $|\gamma|$:

$$|\gamma| < |\sum_j' \frac{1}{\mu_{j,k-2}}| + 2\frac{1}{4\varepsilon} \leq 2\max\{\frac{1}{\lambda_{+,k}}, \frac{1}{|\lambda_{-,k}|}\} + \frac{1}{2\varepsilon} \leq \frac{1}{\varepsilon}, \tag{6.52}$$

where the last inequality follows from (6.49) and the construction of ε. This shows that the only zero of ψ which might belong to $[-4\varepsilon, 4\varepsilon]$, namely $\lambda = 1/\gamma$ remains outside the interval $[-\varepsilon, \varepsilon]$. Thus, the first condition in (6.46) is satisfied for ψ of (6.51). It remains to check the second one. For this to hold, (6.50) turns out to be crucial since it implies that a zero $\mu_{j,k-2}$ in $[-4\varepsilon, 4\varepsilon]$ must at the same time belong to the smaller interval $[-\varepsilon/2, \varepsilon/2]$, and hence,

$$|1 - \frac{\lambda}{\mu_{j,k-2}}| \geq \frac{|\lambda|}{|\mu_{j,k-2}|} - 1 \geq \frac{\varepsilon}{\varepsilon/2} - 1 = 1, \qquad |\lambda| \geq \varepsilon.$$

Consequently, it follows from (6.51) that

$$|\psi(\lambda)| \leq |1 - \gamma\lambda| |p_{k-2}^{[4]}(\lambda)| \leq |\lambda^2 p_{k-2}^{[4]}(\lambda)| \max_{|\lambda|\geq\varepsilon} |\frac{1-\gamma\lambda}{\lambda^2}|, \qquad |\lambda| \geq \varepsilon. \tag{6.53}$$

113

The function $\varphi(\lambda) = \lambda^{-2}(1 - \gamma\lambda)$ has a pole at $\lambda = 0$ and tends to zero as $|\lambda| \to \infty$; besides, it has one local extremum at $\lambda = 2/\gamma$. In view of (6.52) this implies that

$$\max_{|\lambda| \geq \epsilon} |\varphi(\lambda)| = \max\{|\varphi(-\epsilon)|, |\varphi(\epsilon)|, |\varphi(2/\gamma)|\} = \max\{\frac{1 + |\gamma|\epsilon}{\epsilon^2}, \frac{\gamma^2}{4}\} \leq \frac{2}{\epsilon^2}.$$

Inserting this into (6.53) the desired inequality in (6.46) follows.

Final Step. Having now established the assumptions for the second step, (6.47) can be inserted into (6.43) which yields

$$|p_k''(0)|^2 \leq c\,(\omega/\delta)^{4/\mu+1}, \tag{6.54}$$

Thus the assertion (6.40) follows from (6.41) and (6.54).

The convergence of $x_k^\delta \to T^\dagger y$ as $\delta \to 0$, if $y \in \mathcal{D}(T^\dagger)$ does not satisfy Assumption 3.6, can be proved as in Theorem 3.12.   □

The similarity of the above analysis to the one in Section 3.2 suggests the heuristic that, up to a multiplicative constant,

$$\|T^\dagger y - x_k^\delta\| \approx |p_k''(0)|^{1/2}\|y^\delta - Tx_k^\delta\|.$$

As in Section 3.4 this is the motivation for the following heuristic stopping rule:

Stopping Rule 6.16 *Compute*

$$\eta_0 = \eta_1 = \|y^\delta\|, \qquad \eta_k = |p_k''(0)|^{1/2}\|y^\delta - Tx_k^\delta\|, \quad k \geq 2, \tag{6.55}$$

and terminate the MR-II *iteration after* $k(y^\delta)$ *steps, provided* $\eta_{k(y^\delta)} \leq \eta_k$ *for all* $0 \leq k \leq \kappa + 1$.

Note that $p_k''(0)$ is easily updated in the course of the iteration by using the recursion (6.11), i.e.,

$$p_{k+1}''(0) = p_k''(0) - 2\varrho_k u_{k-1}^{[4]}(0).$$

The numbers $u_{k-1}^{[4]}(0)$ in turn can be computed from (6.10):

$$\tilde{u}_{k+1}^{[4]}(0) = -\alpha_k u_k^{[4]}(0) - \beta_k u_{k-1}^{[4]}(0), \quad u_{k+1}^{[4]}(0) = \tilde{u}_{k+1}^{[4]}(0)/\beta_{k+1}, \qquad k \geq 0,$$

starting with $u_{-1}^{[4]}(0) = 0$ and $u_0^{[4]}(0) = 1/\beta_0$.

It is left to the reader to verify the assertions of Theorem 3.14, Corollary 3.15 and Theorem 3.16 for this stopping rule. The proofs from Section 3.4 extend immediately in view of Lemmas 6.12, 6.13, and the estimate (6.54) for $|p_{k(\delta)}''(0)|$ at the stopping index of the discrepancy principle.

6.5 Estimates for the stopping index

It is the aim of the following considerations to provide tools for a comparison of the efficiency of MR-II, mainly as compared to CGNE, but also as compared to MR. All these methods minimize the residual norm in certain Krylov subspaces. The first result is a straightforward consequence of this observation.

Proposition 6.17 *Let T be selfadjoint, indefinite, and denote by k_1 and k_2 the stopping indices of the discrepancy principle (Stopping Rule 3.10) for MR-II and CGNE, respectively. Then one has $2k_1 \leq k_2$. If, in addition, T is semidefinite, and k_0 is the corresponding stopping index of MR, then $k_0 \leq k_1$.*

Proof. Recall that $x_0 = 0$ by convention. By Proposition 2.1, and by the definition of MR-II, the kth iterates x_k^δ of MR (for semidefinite problems only), MR-II, and CGNE, respectively, minimize the residual $\|y^\delta - Tx_k^\delta\|$ in the Krylov subspaces

$$\mathcal{K}_0(k) = \mathcal{K}_{k-1}(y; T), \quad \mathcal{K}_1(k) = \mathcal{K}_{k-2}(Ty; T), \quad \mathcal{K}_2(k) = \mathcal{K}_{k-1}(Ty; T^2).$$

Since

$$\mathcal{K}_1(k) \subset \mathcal{K}_0(k) \quad \text{and} \quad \mathcal{K}_2(k) \subset \mathcal{K}_1(2k),$$

the claim follows from the definition of the stopping rule. $\square$

This allows the following important conclusion: as MR-II requires at most twice as many iterations than CGNE, with each MR-II iteration being approximately half as expensive as one CGNE iteration (at least as far as multiplications with T are concerned), MR-II seems to be the method of choice for indefinite problems.

Since quantitative estimates for $k(\delta)$ are known for CGNE (cf. the final remark in Section 5.1), Proposition 6.17 can not only be used for comparison, but also to obtain bounds for the stopping index $k(\delta)$ of MR-II.

Corollary 6.18 *Let $\|y - y^\delta\| \leq \delta$ and $k(\delta)$ be the stopping index for MR-II as determined by Stopping Rule 3.10. If y satisfies Assumption 3.6 then*

$$k(\delta) \leq c \, (\omega/\delta)^{1/\mu+1}.$$

If, in addition, T is a non-degenerate compact operator with eigenvalues $|\lambda_j| = O(j^{-\alpha})$ for some $\alpha > 0$ and $j \to \infty$, then

$$k(\delta) \leq c \, (\omega/\delta)^{1/(\mu+1)(\alpha+1)}.$$

If the eigenvalues of T decay in absolute value like $O(q^j)$ as $j \to \infty$ with $q < 1$ then

$$k(\delta) \leq c \, (1 + \log^+(\omega/\delta)).$$

The remainder of this section is concerned with the question what conditions are in favor of MR-II as compared to CGNE. To this end consider the following two quite general case studies.

Example 6.19 Let $d\|E_\lambda y^\delta\|^2$ be symmetric with respect to the origin. Clearly, this implies that $\lambda^2 d\|E_\lambda y^\delta\|^2$ is also symmetric. It is known (compare [9, Theorem 1.4.3]) that the corresponding orthogonal polynomials $u_k^{[2]}$ are even for k even, and odd for k odd. Since all roots of $u_k^{[2]}$ are simple it follows that

$$u_{2k}^{[2]}(0) \neq 0, \qquad u_{2k}^{[2]\prime}(0) = 0, \qquad k \in \mathbb{N}_0.$$

Consequently, for every $k \in \mathbb{N}_0$ there is a multiple $p_{2k}^{[2]} \in \Pi_{2k}^{00}$ of $u_{2k}^{[2]}$, which satisfies

$$[p_{2k}^{[2]}, p]_2 = 0 \qquad \text{for every } p \in \Pi_{2k-1}.$$

Hence, by virtue of Proposition 6.1, the residual polynomials $\{p_k\}$ of MR-II are given by

$$p_{2k} = p_{2k+1} = p_{2k}^{[2]}, \qquad k \in \mathbb{N}_0.$$

In other words, for every $k \in \mathbb{N}_0$ the two consecutive iterates x_{2k} and x_{2k+1} coincide, and the iteration polynomials $q_{k-1} = (1 - p_k)/\lambda$ are all odd. This shows that

$$x_{2k}, x_{2k+1} \in \mathcal{K}_{k-1}(Ty; T^2),$$

and hence, x_{2k} and x_{2k+1} coincide with the kth iterate of CGNE. In other words, in this situation the upper bound for the stopping index of MR-II as given in Proposition 6.17 is attained.

The second example considers the situation where T is almost positive definite.

Example 6.20 Assume that y satisfies Assumption 3.6, and T has just finitely many negative eigenvalues

$$\lambda_1 < \lambda_2 < \ldots < \lambda_r < 0.$$

Here, r may be any nonnegative integer; in particular, when $r = 0$ then T is semidefinite, and one may be interested in comparing MR-II with MR. Let

$$\psi_r(\lambda) = \prod_{j=1}^{r} (1 - \frac{\lambda}{\lambda_j}),$$

and define $\varphi_{k-r}^{[0]}$ by (5.2) as a Jacobi polynomial of degree $k - r$ with $\nu = 2\mu + 2$; similarly, let $\varphi_{k-r-1}^{[2]}$ be the corresponding Jacobi polynomial (5.2) of degree $k - r - 1$ with $\nu = 2\mu + 4$. Consider the following comparison polynomial of degree k:

$$p = \psi_r(\varphi_{k-r}^{[0]} - \gamma_k \lambda \varphi_{k-r-1}^{[2]}). \tag{6.56}$$

116

Here, γ_k is chosen so as to have $p \in \Pi_k^{00}$, i.e.,

$$\gamma_k = \psi_r'(0) + \varphi_{k-r}^{[0]}{}'(0) \,.$$

Consequently, as shown in Lemma 4.6, (iii),

$$|\gamma_k| \sim k^2, \qquad k \to \infty \,. \tag{6.57}$$

Since ψ_r is bounded on $[0, 1]$, it follows that

$$|\lambda^{\mu+1} p(\lambda)| \leq c \|\lambda^{\mu+1} \varphi_{k-r}^{[0]}\|_{[0,1]} + c\, |\gamma_k|\, \|\lambda^{\mu+2} \varphi_{k-r-1}^{[2]}\|_{[0,1]}, \qquad 0 \leq \lambda \leq 1 \,,$$

and inserting (5.3) and (6.57) this yields

$$|\lambda^{\mu+1} p(\lambda)| \leq c(k+1)^{-2\mu-2}, \qquad 0 \leq \lambda \leq 1 \,. \tag{6.58}$$

Since p vanishes on all negative eigenvalues of T, it follows that

$$\|p(T)y\| = \|p(T)T^{\mu+1}w\| \leq c(k+1)^{-2\mu-2}\omega \,.$$

It is not difficult to see that (6.58) implies a uniform upper bound for p on $[0,1]$, independent of k, although this bound need not be 1 as in all former examples. Nevertheless, this implies that after approximately $(\omega/\delta)^{1/2\mu+2}$ iterations the residual $\|y^\delta - Tx_k^\delta\|$ of MR-II will already be of the order of δ. Note that this is about the same estimate as has been obtained for MR in Theorem 5.1. Consequently, when T is semidefinite then one may expect similar error histories for MR and MR-II.

The conclusion is the following. If the spectral mass of y^δ splits into similar components corresponding to negative and positive parts of the spectrum of T then little gain may be expected from using MR-II instead of CGNE. Here, "similar components" refers to approximately equal mass, equal clustering properties of eigenvalues, and equal asymptotics for the decay of spectral contributions near both sides of the origin. See Section 6.7 for such an example; the numerical experiments nevertheless show a substantial improvement by using MR-II instead of CGNE.

On the other hand, if a dominating portion of the spectral mass is located on one side of the origin then MR-II may be up to an order of magnitude faster than CGNE. For certain spectral distributions one can expect about the same improvement as when using MR instead of CGNE for semidefinite problems. To some extent this can be exemplified with the image reconstruction problem of Section 5.3. Although the continuous problem is positive definite, the discretized problem is symmetric indefinite. However, being an approximation for a definite problem, the negative spectral elements only contribute to some minor extent. Numerical examples for this problem are given in the following section.

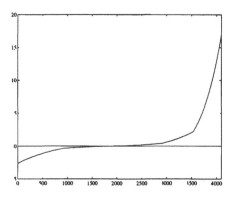

Fig. 6.2. Eigenvalues of A

6.6 The image reconstruction problem

Consider the deconvolution problem from Section 5.3. The operator T in (5.16) is self-adjoint because of the symmetry of the kernel function h of (5.17), and the spectrum of T is given by the values of the Fourier transform of h, namely by the values of

$$\hat{h}(\sigma, \tau) = \frac{\pi}{\chi} \exp(-\frac{1}{4\chi}(\sigma^2 + \tau^2))$$

over $\mathbf{R}^2$. It follows that T is positive definite and $\|T\| = \pi/\chi$.

Although the symmetry is maintained in the simple discretization of Section 5.3, the resulting matrix A of (5.18) is indefinite, cf. Figure 6.2. The reason is that the spectrum of the discretized operator is more related to the oscillating Fourier transform of the truncated point spread function rather than to the Gaussian presented above.

The discrete problem (5.18) is a first example of a selfadjoint and indefinite problem. Note that the distribution of the eigenvalues is not symmetric. Although most of the spectrum is within an interval $(-2.66, 2.66)$, there are about 500 out of the 4096 eigenvalues of A which dominate the positive part of the spectrum, cf. Figure 6.2. One may therefore expect that the spectral contribution coming from the positive eigenvalues dominates the contribution from the negative eigenvalues, which is likely to be in favor of MR-II.

This is confirmed by the numerical results for MR-II and CGNE shown in Tables 6.1 and 6.2 which correspond to exactly the same experiments as in Section 5.3. Recall that the entries of the two tables represent averages over all twenty runs; as in Section 5.3 the individual results agree very well with these averages. It can nicely be seen that the numbers in Table 6.1 are almost the same for both methods (whether

118

	1% noise			0.1% noise		
	opt.	ord.	heu.	opt.	ord.	heu.
MR-II	0.1454	0.1610	0.2452	0.0755	0.0909	0.0811
CGNE	0.1452	0.1606	0.2514	0.0753	0.0915	0.0811

Table 6.1. Average (relative) error norm

	1% noise			0.1% noise		
	opt.	ord.	heu.	opt.	ord.	heu.
MR-II	20.1	9.0	2.0	132.8	54.2	77.2
CGNE	26.3	13.0	2.0	166.1	68.6	96.8

Table 6.2. Average iteration count

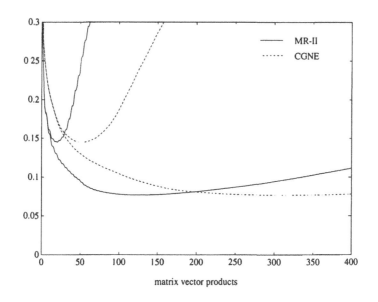

Fig. 6.3. Iteration error for MR-II and CGNE

they correspond to the optimal stopping index or to one of the stopping rules). On the other hand, the iteration counts listed in Table 6.2 are significantly better for MR-II. The improvement from about 166 to 133 iterations in the case of 0.1% noise may not seem dramatic on the first glimpse, however, in view of the fact that one CGNE iteration requires two matrix vector multiplications instead of one, the use of MR-II results in a speedup of more than a factor of two. This is illuminated by Figure 6.3 which contains a plot of the error history for two particular right-hand sides (with 1% and 0.1% noise, respectively) versus the number of matrix vector multiplications.

6.7 The sideways heat equation

The second test example is concerned with determining an inaccessible surface temperature of a body given measurements of the temperature inside the body. Physical examples include the determination of the surface temperature of a space vehicle reentering the earth's atmosphere, and the determination of the temperature within a reactor from thermocouples buried inside the wall, close to the wall's inner surface.

The numerical experiments that are described in the sequel deal with a simple 1D model problem, namely

$$
\begin{aligned}
u_t &= u_{ss}, & 0 < s < \infty,\ 0 < t < \infty, \\
u(s,0) &= 0, & 0 \le s \le \infty,
\end{aligned}
$$

where the temperature $y = u(1, \cdot) \in \mathcal{L}^2(\mathbb{R}^+)$ is known (i.e., measured), and the temperature $u(0, t)$, $t > 0$, at the boundary is sought. In order to guarantee well-posedness of the corresponding direct problem (where the boundary temperature is given and $u(1, t)$ is sought) it is assumed that the temperature u remains bounded in the quarter plane $s > 0$, $t > 0$, cf., e.g., [8, Chapter 4].

The solution $x = u(0, \cdot)$ of the above sideways heat equation can be determined from a Volterra convolution equation of the first kind, cf. [8],

$$
y(t) = (Kx)(t) = \int_0^t h(t - t')x(t')dt', \tag{6.59}
$$

with kernel function

$$
h(t) = \frac{1}{2\sqrt{\pi}} t^{-3/2} \exp(-\frac{1}{4t}), \qquad 0 < t \le 1.
$$

The interesting feature of this example is the fact that K is a compact operator with rapidly decreasing singular values because h vanishes exponentially at the origin.

Assume that there are given n discrete measurements $y(t_m)$ at equidistant time steps $t_m \in [0, 1]$, $1 \le m \le n$. For simplicity let n be even. Simple collocation with piecewise linear splines leads to a nonsingular linear system $Ax = y$ with a lower triangular Toeplitz matrix A. In principle one could do a simple forward solve to compute $\mathbf{x}$ from $\mathbf{y}$ but this is an extremely unstable procedure since the diagonal pivot elements of A are given by $h(t_1/2)$, and hence, these pivots are almost numerically zero because of the very slow increase of h near the origin.

Although A is not symmetric, it can easily be transformed into a symmetric matrix by flipping A in the up/down direction. The resulting matrix H is a Hankel matrix, i.e., constant along counter-diagonals, with a large zero block of dimension $n/2 \times n/2$ in the lower right corner. As a matter of fact, H is indefinite with precisely $n/2$ positive and negative eigenvalues, respectively, cf. [44]. It is easy to see that the eigenvalues of H coincide with the singular values of A in absolute value.

Permuting $\mathbf{y}$ in the same way (which gives $\mathbf{b}$, say), the vector $\mathbf{x}$ solves the linear system

$$H\mathbf{x} = \mathbf{b}, \tag{6.60}$$

and MR-II can be used for computing regularized approximations of $\mathbf{x}$. Since $H\mathbf{x}$ is just a permutation of $A\mathbf{x}$ it is clear that one could apply FFT techniques for implementing multiplications by H with only $O(n\cdot\log n)$ operations, cf. [38]; for smaller values of n standard matrix vector multiplication is faster, though.

In the present example, taken from [15] and implemented in routine **heat** of [41], x is piecewise continous,

$$
x(t) = \begin{cases}
\frac{300}{4} t^2, & 0 \leq t \leq 1/10, \\
\frac{3}{4} + (20t - 2)(3 - 20t), & 1/10 < t \leq 3/20, \\
\frac{3}{4} e^{2(3-20t)}, & 3/20 < t \leq 1/2, \\
0, & 1/2 < t \leq 1.
\end{cases}
$$

The dimension of H is 128×128. The mass points of the discrete inner product $[\cdot,\cdot]_0$ are shown in Figure 6.4: each line represents an eigenvector component of $\mathbf{b}$ at an eigenvalue of H. Essential parts of the total spectral mass are on both sides of the origin, although the spectrum of H itself is not symmetric.

The experimental setting, i.e., generation of noise, number of experiments, and so forth, is the same as in Section 5.3. Thus, there are twenty experiments for two different noise levels, each, with the resulting average errors and iteration counts displayed in Tables 6.3 and 6.4. As in the example from the previous section, the best approximations by MR-II and CGNE have about the same accuracy. In fact they are very close themselves as can be seen from Figure 6.5, in particular when looking at the zoomed details in the right-hand side plot.

The stopping rules perform slightly better for CGNE. In view of the discussion in Section 6.5 it is somewhat surprising that the iteration counts for MR-II and CGNE are almost the same. This is probably due to the fact that the eigenvalues are not completely symmetric with respect to the origin, and therefore H^2 has about the same number of isolated dominating eigenvalues as H. This observation is somehow

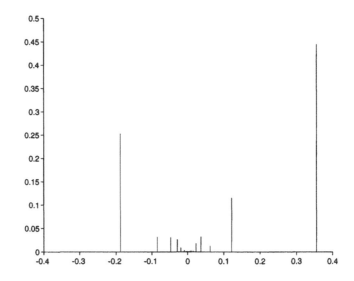

Fig. 6.4. Spectral masses for inner product $[\cdot, \cdot]_0$

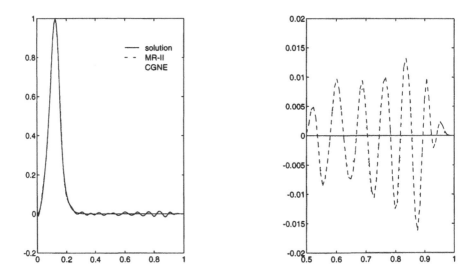

Fig. 6.5. Exact solution and its reconstructions, with differing scales

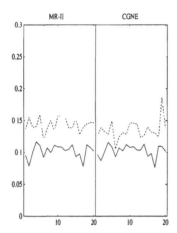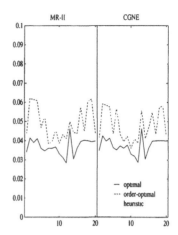

Fig. 6.6. Relative errors of the methods for 1% noise (left) and 0.1% noise (right)

supported by the technique of proof for Theorem 5.2 concerning the estimation of the stopping index $k(\delta)$ for compact operator equations.

As a consequence, CGNE is about twice as expensive as MR-II for this example; this is illustrated by Figure 6.7 where the relative error history is plotted versus the number of matrix vector products in the case of 0.1% noise.

Finally, it is worthwhile to comment on the differences to the image reconstruction example where the underlying operator has not been compact. It is obvious that far less iterations are required for the present example, and the accuracy is nevertheless better. Although the degree of smoothness of the true solution (as measured by Assumption 3.6) is not known in either example, the numbers support to some extent the result that conjugate gradient type methods need fewer iterations for compact operator equations.

In the sideways heat equation problem the heuristic stopping rules for CGNE and MR-II lead to approximations which are worse than the optimal approximations by a factor of about two, whereas the order-optimal stopping rules are somewhere in between. This is worse than in the image reconstruction problem. Nonetheless, as can be seen from Figure 6.8, the error approximations η_k on which the heuristic stopping rule is based, cf. (6.55), follow the true errors fairly well. In fact, the stopping indices for the heuristic rules were never that unrealistic as in the third column of Table 6.2.

Summarizing, MR-II is a more efficient method for indefinite problems, even when the spectral measure seems to be fairly symmetric with respect to the origin. Properly interpreted, both stopping rules give reasonable results. The heuristic rule is less robust, though, and its performance strongly depends on the underlying problem. Of

	1% noise			0.1% noise		
	opt.	ord.	heu.	opt.	ord.	heu.
MR-II	0.1022	0.1439	0.2158	0.0369	0.0489	0.0725
CGNE	0.1026	0.1350	0.2035	0.0373	0.0478	0.0713

Table 6.3. Average (relative) error norm

	1% noise			0.1% noise		
	opt.	ord.	heu.	opt.	ord.	heu.
MR-II	15.7	11.0	9.0	28.4	20.7	17.9
CGNE	13.9	10.9	8.0	27.9	20.2	16.0

Table 6.4. Average iteration count

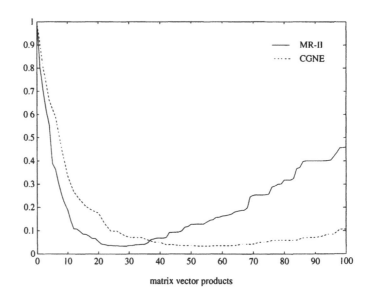

Fig. 6.7. Iteration error of MR-II and CGNE for 0.1% noise

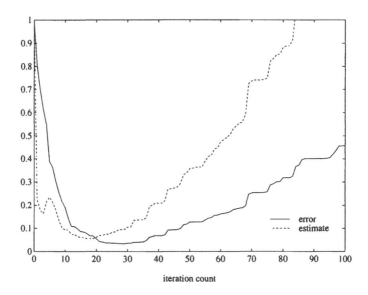

Fig. 6.8. Iteration error of MR-II and heuristic error estimate for 0.1% noise

course, this should be expected as it is primarily based on heuristic assumptions which in some examples may make more sense than in others.

Notes and remarks

Section 6.1. The original implementations of MR and MR-II for indefinite problems were developed by PAIGE and SAUNDERS [64]. The ORTHODIR implementation of MR is described in more detail in HACKBUSCH [32, Sect. 9.5.3] and in FISCHER [18]. The name ORTHODIR was introduced by YOUNG and JEA [84].

The idea of choosing approximations from the Krylov subspace (6.2) originates with a paper by FRIDMAN [21], and has been further recommended in [64] in the context of singular algebraic equations with incompatible right hand side. The MR-II implementation in [64] is significantly different from Algorithm 6.1. The presentation as given here follows [19].

As mentioned right above, the use of residual polynomials p_k with $p'_k(0) = 0$ has some tradition in the context of singular linear matrix equations. This is so because "incompatible" components of the right-hand side are amplified during the iteration by $-p'_k(0)$, which typically diverges to infinity. Hence, stipulating $p'_k(0) = 0$ provides a way to prevent divergence; compare [19] and references therein.

Section 6.2. Interlacing properties for the roots of linear combinations of two consecutive orthogonal polynomials are classical, cf. SZEGÖ [76, Theorem 3.3.4]. Linear combinations of three consecutive orthogonal polynomials were studied by MICCHELLI and RIVLIN [55]. The proof of Proposition 6.6 uses the technique of [55].

Section 6.3. It remains an open problem whether the iteration error of MR-II decays monotonically for $y \in \mathcal{D}(T^\dagger)$. It is for this reason that the proofs of Theorem 6.9 and Theorem 3.4 are completely different. The proof of Theorem 6.9 is more related to the convergence analysis of MR as given by NEMIROVSKII and POLYAK [59].

Section 6.4. O'LEARY and SIMMONS [61] mention the possibility of regularizing indefinite problems by Lanczos tridiagonalization, which would be equivalent to the MR method. However, there is no theoretical justification of this approach in the literature.

Section 6.5. Similar estimates have been given by FREUND [20] for conjugate gradient type methods for well-posed indefinite problems.

Section 6.7. ELDÉN's paper [15], where this particular example $x(t)$ has been used for numerical experiments, contains a nice survey of regularization techniques for the sideways heat equation. The idea to transform the problem into a linear system with a Hankel matrix is taken from a paper by EKSTROM and RHOADS [14], see also O'LEARY and SIMMONS [61].

126

References

[1] O. M. ALIFANOV AND S. V. RUMJANCEV, *On the stability of iterative methods for the solution of linear ill-posed problems*, Soviet Math. Dokl., 20 (1979), pp. 1133–1136.

[2] G. ANGER, R. GORENFLO, H. JOCHMANN, H. MORITZ, AND W. WEBERS, eds., *Inverse Problems: Principles and Applications in Geophysics, Technology and Medicine*, Berlin, 1993, Akademie Verlag.

[3] R. ASKEY, *Orthogonal expansions with positive coefficients*, Proc. Amer. Math. Soc., 16 (1965), pp. 1191–1194.

[4] A. B. BAKUSHINSKII, *Remarks on choosing a regularization parameter using the quasi-optimality and ratio criterion*, USSR Comput. Math. and Math. Phys., 24,4 (1984), pp. 181–182.

[5] J. BAUMEISTER, *Stable Solution of Inverse Problems*, Friedr. Vieweg & Sohn, Braunschweig, 1987.

[6] H. BRAKHAGE, *On ill-posed problems and the method of conjugate gradients*, in Inverse and Ill-Posed Problems, H. W. Engl and C. W. Groetsch, eds., Boston, New York, London, 1987, Academic Press, pp. 165–175.

[7] D. CALVETTI, L. REICHEL, AND Q. ZHANG, *Conjugate gradient algorithms for symmetric inconsistent linear systems*, in Proceedings of the Lanczos International Centenary Conference, J. D. Brown, M. T. Chu, D. C. Ellis, and R. J. Plemmons, eds., Philadelphia, 1994, SIAM, pp. 267–272.

[8] J. R. CANNON, *The One-Dimensional Heat Equation*, Addison-Wesley, Reading, MA, 1984.

[9] T. S. CHIHARA, *An Introduction to Orthogonal Polynomials*, Gordon and Breach, Science Publishers, New York, 1978.

[10] ——, *Orthogonal polynomials and measures with end point masses*, Rocky Mountain J. Math., 15 (1985), pp. 705–719.

[11] E. J. CRAIG, *The n-step iteration procedures*, J. Math. Phys., 34 (1955), pp. 64–73.

[12] V. DRUSKIN AND L. KNIZHNERMAN, *Error bounds in the simple Lanczos procedure for computing functions of symmetric matrices and eigenvalues*, Comput. Math. Math. Phys., 31,7 (1991), pp. 20–30.

[13] B. EICKE, A. K. LOUIS, AND R. PLATO, *The instability of some gradient methods for ill-posed problems*, Numer. Math., 58 (1990), pp. 129–134.

[14] M. P. EKSTROM AND R. L. RHOADS, *On the application of eigenvector expansions to numerical deconvolution*, J. Comput. Phys., 14 (1974), pp. 319–340.

[15] L. ELDÉN, *Numerical solution of the sideways heat equation*, to appear in [17].

[16] H. W. ENGL, *Regularization methods for the stable solution of inverse problems*, Surveys Math. Indust., 3 (1993), pp. 71-143.

[17] H. W. ENGL AND W. RUNDELL, eds., *Inverse Problems in Diffusion Processes*, Philadelphia, 1995, SIAM, to appear.

[18] B. FISCHER, *Orthogonal polynomials and polynomial based iteration methods for indefinite linear systems*, Habilitationsschrift, Universität Hamburg, Hamburg, 1993.

[19] B. FISCHER, M. HANKE, AND M. HOCHBRUCK, *A note on conjugate-gradient type methods for indefinite and/or inconsistent linear systems*, submitted.

[20] R. FREUND, *Über einige CG-ähnliche Verfahren zur Lösung linearer Gleichungssysteme*, Dissertation, Universität Würzburg, Würzburg, 1984.

[21] V. M. FRIDMAN, *The method of minimum iterations with minimum errors for a system of linear algebraic equations with a symmetrical matrix*, USSR Comput. Math. and Math. Phys., 2 (1963), pp. 362–363.

[22] S. F. GILYAZOV, *Iterative solution methods for inconsistent operator equations*, Moscow Univ. Comput. Math. Cybernet., 3 (1977), pp. 78–84.

[23] ——, *Regularizing algorithms based on the conjugate gradient method*, USSR Comput. Math. and Math. Phys., 26,1 (1986), pp. 8–13.

[24] ——, *Methods for solving linear ill-posed problems*, Moskov. Gos. Univ., Moscow, 1987. In Russian.

[25] G. H. GOLUB AND D. P. O'LEARY, *Some history of the conjugate gradient and Lanczos algorithms: 1948-1976*, SIAM Rev., 31 (1989), pp. 50-102.

[26] G. H. GOLUB AND C. F. VAN LOAN, *Matrix Computations*, The Johns Hopkins University Press, Baltimore, London, 1989.

[27] A. GREENBAUM, *Behavior of slightly perturbed Lanczos and conjugate-gradient recurrences*, Linear Algebra Appl., 113 (1989), pp. 7-63.

[28] A. GREENBAUM AND Z. STRAKOŠ, *Predicting the behavior of finite precision Lanczos and conjugate gradient computations*, SIAM J. Matrix Anal. Appl., 13 (1992), pp. 121-137.

[29] C. W. GROETSCH, *Generalized Inverses of Linear Operators*, Marcel Dekker, Inc., New York, Basel, 1977.

[30] ——, *The Theory of Tikhonov Regularization for Fredholm Equations of the First Kind*, Pitman Publishing, Boston, London, Melbourne, 1984.

[31] M. H. GUTKNECHT, *Changing the norm in conjugate gradient type algorithms*, SIAM J. Numer. Anal., 30 (1993), pp. 40-56.

[32] W. HACKBUSCH, *Iterative Solution of Large Sparse Systems of Equations*, Springer-Verlag, New York, Berlin, Heidelberg, 1994.

[33] J. HADAMARD, *Lectures on Cauchy's Problem in Linear Partial Differential Equations*, Yale University Press, New Haven, 1923.

[34] M. HANKE, *Accelerated Landweber iterations for the solution of ill-posed equations*, Numer. Math., 60 (1991), pp. 341-373.

[35] ———, *The minimal error conjugate gradient method is a regularization method*, to appear in Proc. Amer. Math. Soc.

[36] ———, *Asymptotics of orthogonal polynomials and the numerical solution of ill-posed problems*, to appear in Ann. Numer. Math.

[37] M. HANKE AND H. W. ENGL, *An optimal stopping rule for the ν-method for solving ill-posed problems using Christoffel functions*, J. Approx. Theory, 79 (1994), pp. 89-108.

[38] M. HANKE AND P. C. HANSEN, *Regularization methods for large-scale problems*, Surveys Math. Indust., 3 (1993), pp. 253-315.

[39] M. HANKE, J. G. NAGY, AND R. J. PLEMMONS, *Preconditioned iterative regularization for ill-posed problems*, in Numerical Linear Algebra, L. Reichel, A. Ruttan, and R. S. Varga, eds., Berlin, New York, 1993, de Gruyter, pp. 141-163.

[40] M. HANKE AND T. RAUS, *A general heuristic for choosing the regularization parameter in ill-posed problems*, submitted.

[41] P. C. HANSEN, REGULARIZATION TOOLS: *A* MATLAB *package for analysis and solution of discrete ill-posed problems*, Numer. Algorithms, 6 (1993), pp. 1-35.

[42] ———, *Experience with regularizing CG iterations*, to appear in BIT.

[43] R. M. HAYES, *Iterative methods of solving linear problems on Hilbert space*, in Contributions to the Solution of Systems of Linear Equations and the Determination of Eigenvalues, O. Taussky, ed., Washington, 1954, Nat. Bur. Standards. Appl. Math. Ser. 39, pp. 71-103.

[44] E. V. HAYNSWORTH AND A. M. OSTROWSKI, *On the inertia of some classes of partitioned matrices*, Linear Algebra Appl., 1 (1968), pp. 299-316.

[45] M. R. HESTENES AND E. STIEFEL, *Methods of conjugate gradients for solving linear systems*, J. Research Nat. Bur. Standards, 49 (1952), pp. 409-436.

[46] W. J. KAMMERER AND M. Z. NASHED, *On the convergence of the conjugate gradient method for singular linear operator equations*, SIAM J. Numer. Anal., 9 (1972), pp. 165-181.

[47] J. T. KING, *A minimal error conjugate gradient method for ill-posed problems*, J. Optim. Theory Appl., 60 (1989), pp. 297-304.

[48] T. H. KOORNWINDER, *Orthogonal polynomials with weight function $(1-x)^\alpha (1+x)^\beta + m\delta(x+1) + n\delta(x-1)$*, Canad. Math. Bull., 27 (1984), pp. 205-214.

[49] R. L. LAGENDIJK AND J. BIEMOND, *Iterative Identification and Restoration of Images*, Kluwer, Boston, Dordrecht, London, 1991.

[50] C. LANCZOS, *Solution of systems of linear equations by minimized iterations*, J. Research Nat. Bur. Standards, 49 (1952), pp. 33-53.

[51] L. J. LARDY, *A class of iterative methods of conjugate gradient type*, Numer. Funct. Anal. Optim., 11 (1990), pp. 283-302.

[52] A. S. LEONOV, *On the choice of regularization parameters by means of the quasi-optimality and ratio criteria*, Soviet Math. Dokl., 19 (1978), pp. 537-540.

[53] A. K. LOUIS, *Convergence of the conjugate gradient method for compact operators*, in Inverse and Ill-Posed Problems, H. W. Engl and C. W. Groetsch, eds., Boston, New York, London, Tokyo, Toronto, 1987, Academic Press, pp. 177-183.

[54] ——, *Inverse und schlecht gestellte Probleme*, B.G. Teubner, Stuttgart, 1989.

[55] C. A. MICCHELLI AND T. J. RIVLIN, *Numerical integration rules near Gaussian quadrature*, Israel J. Math., 16 (1973), pp. 287-299.

[56] V. A. MOROZOV, *On the solution of functional equations by the method of regularization*, Soviet Math. Dokl., 7 (1966), pp. 414-417.

[57] F. NATTERER, *The Mathematics of Computerized Tomography*, Wiley, Chichester, New York, Brisbane, Toronto, Singapore, 1986.

[58] A. S. NEMIROVSKII, *The regularization properties of the adjoint gradient method in ill-posed problems*, USSR Comput. Math. and Math. Phys., 26,2 (1986), pp. 7-16.

[59] A. S. NEMIROVSKII AND B. T. POLYAK, *Iterative methods for solving linear ill-posed problems under precise information I*, Engrg. Cybernetics, 22,3 (1984), pp. 1-11.

[60] Y. NOTAY, *On the convergence rate of the conjugate gradients in presence of rounding errors*, Numer. Math., 65 (1993), pp. 301-317.

[61] D. P. O'LEARY AND J. A. SIMMONS, *A bidiagonalization-regularization procedure for large scale discretizations of ill-posed problems*, SIAM J. Sci. Statist. Comput., 2 (1981), pp. 474-489.

[62] C. C. PAIGE, *Bidiagonalization of matrices and solution of linear equations*, SIAM J. Numer. Anal., 11 (1974), pp. 197-209.

[63] ——, *Error analysis of the Lanczos algorithm for tridiagonalizing a symmetric matrix*, J. Inst. Math. Appl., 18 (1976), pp. 341-349.

[64] C. C. PAIGE AND M. A. SAUNDERS, *Solution of sparse indefinite systems of linear equations*, SIAM J. Numer. Anal., 12 (1975), pp. 617-629.

[65] ——, *LSQR: an algorithm for sparse linear equations and sparse least squares*, ACM Trans. Math. Software, 8 (1982), pp. 43-71.

[66] L. PÄIVÄRINTA AND E. SOMERSALO, eds., *Inverse Problems in Mathematical Physics*, Berlin, Heidelberg, 1993, Springer.

[67] R. PLATO, *Optimal algorithms for linear ill-posed problems yield regularization methods*, Numer. Funct. Anal. Optim., 11 (1990), pp. 111-118.

[68] ——, *Über die Diskretisierung und Regularisierung schlecht gestellter Probleme*, Dissertation, TU Berlin, Berlin, 1990.

[69] T. RAUS, *The principle of the residual in the solution of ill-posed problems with non-selfadjoint operator*, Tartu Riikl. Ül. Toimetised, 715 (1985), pp. 12-20. In Russian.

[70] J. B. READE, *Eigenvalues of positive definite kernels*, SIAM J. Math. Anal., 14 (1983), pp. 152-157.

[71] F. RIESZ AND B. SZ.-NAGY, *Functional Analysis*, Ungar, New York, 1955.

[72] P. C. SABATIER, ed., *Inverse Methods in Action*, Berlin, Heidelberg, New York, 1990, Springer Verlag.

[73] V. E. SHAMANSKII, *On certain numerical schemes for iteration processes*, Ukrain. Mat. Zh., 14 (1962), pp. 100-109. In Russian.

[74] E. STIEFEL, *Relaxationsmethoden bester Strategie zur Lösung linearer Gleichungssysteme*, Comment. Math. Helv., 29 (1955), pp. 157-179.

[75] Z. STRAKOŠ, *On the real convergence rate of the conjugate gradient method*, Linear Algebra Appl., 154-156 (1991), pp. 535-549.

[76] G. SZEGÖ, *Orthogonal Polynomials*, Amer. Math. Soc. Colloq. Publ., Vol. 23, Amer. Math. Soc., Providence, Rhode Island, 1975.

[77] A. N. TIKHONOV, *Solution of incorrectly formulated problems and the regularization method*, Soviet Math. Dokl., 4 (1963), pp. 1035-1038.

[78] A. N. TIKHONOV AND V. Y. ARSENIN, *Solutions of Ill-Posed Problems*, John Wiley & Sons, New York, Toronto, London, 1977.

[79] A. N. TIKHONOV, A. S. LEONOV, A. I. PRILEPKO, I. A. VASIN, V. A. VATUTIN, AND A. G. YAGOLA, eds., *Ill-Posed Problems in Natural Sciences*, Utrecht, Moscow, 1992, VSP BV/TVP Sci. Publ.

[80] W. F. TRENCH, *Proof of a conjecture of Askey on orthogonal expansions with positive coefficients*, Bull. Amer. Math. Soc., 81 (1975), pp. 954-956.

[81] H. TRIEBEL, *Interpolation Theory, Function Spaces, Differential Operators*, North-Holland, Amsterdam, New York, Oxford, 1978.

[82] G. M. VAINIKKO AND A. Y. VERETENNIKOV, *Iteration Procedures in Ill-Posed Problems*, Nauka, Moscow, 1986. In Russian.

[83] M. YAMAGUTI, K. HAYAKAWA, Y. ISO, M. MORI, T. NISHIDA, K. TOMOEDA, AND M. YAMAMOTO, eds., *Inverse Problems in Engineering Sciences*, Tokyo, Berlin, Heidelberg, 1991, Springer Verlag.

[84] D. M. YOUNG AND K. C. JEA, *Generalized conjugate-gradient acceleration of nonsymmetric iterative methods*, Linear Algebra Appl., 34 (1980), pp. 159-194.

Index

Printed and bound by CPI Group (UK) Ltd, Croydon, CR0 4YY

23/10/2024

01778230-0006